Newton Rules Biology

Newton Rules Biology

A Physical Approach to Biological Problems

C.J. PENNYCUICK

Maytag Professor of Ornithology
University of Miami

OXFORD   NEW YORK   TOKYO
OXFORD UNIVERSITY PRESS
1992

Oxford University Press, Walton Street, Oxford OX2 6DP
Oxford   New York   Toronto
Delhi   Bombay   Calcutta   Madras   Karachi
Petaling Jaya   Singapore   Hong Kong   Tokyo
Nairobi   Dar es Salaam   Cape Town
Melbourne   Auckland
and associated companies in
Berlin   Ibadan

Oxford is a trade mark of Oxford University Press

Published in the United States
by Oxford University Press, New York

A catalogue record for this book is available from the British Library

Library of Congress Cataloging in Publication Data
Pennycuick, C. J. (Colin J.)
Newton rules biololgy : a physical approach to biological
problems / C. J. Pennycuick.
Includes bibliographical references and index.
1. Biomechanics.   2. Biophysics.   3. Animal mechanics.
4. Ecology.   5. Animals—physiology.   I. Title.
[DNLM: 1. Biomechanics.   2. Biophysics.   3. Environment.   QT 34 P416n]
QH513.P46   1992   591.1—dc20   91-39081

ISBN 0-19-854020-5 (hbk.)
ISBN 0-19-854021-3 (pbk.)

Set by Colset (Private) Ltd
Printed in Great Britain by
Dotesios Ltd, Trowbridge, Wilts

Preface

The newtonian approach to biology is not new, but has always been a minority interest. Among the multitudes of biochemists and molecular biologists, those who study biomechanics are considered to be the lunatic fringe. Very few among this band of eccentrics have commanded universal respect, and foremost of those was the legendary physiologist A.V. Hill. I hesitate to claim too direct a connection with Hill's work, for fear of tarnishing his memory with my errors, but it has been one of my objectives in writing this book to show how Hill's way of thinking created a thread which links biological events at the cellular level, through animal locomotion, to the large-scale properties of ecosystems.

Hill's scientific style is very unlike that usually associated with the science of physiology. The reader is never in any doubt about the exact physical nature of the quantities he measures, or of the meaning of the units with which he measures them. This eccentricity is traceable to his unusual background. Uniquely for someone who became a physiologist in the early 1900s, Hill had no training in medicine, but started out as a mathematician. His bent was not towards elaborate mathematical modelling, but rather for thinking about biological processes in newtonian terms. Hill's main interest was in the mechanical, thermal, and chemical events in muscles, at the level of the muscle cell and the whole muscle. Most physiologists would consider his 1938 paper, in which he formulated 'Hill's equation', as his most famous work. However, in my view, his 1950 paper *The dimensions of animals and their muscular dynamics* is a classic of equal if not greater stature. This paper originated as an Evening Discourse at the Royal Institution in 1949, and Hill seems to have intended it more as an entertainment than as a serious contribution to science. Amusing it is, but it also started a whole new line of thinking in biology. Although the notion of scale effects in biology was already well known from the work of D'Arcy Thompson (who himself acknowledged that Galileo and Aristotle had anticipated him), Hill was the one who set dimensional reasoning in motion, so to speak, by applying it to quantities like frequency, work, and power.

Hill's method of reasoning in the 1950 paper has its roots in engineering, but is unlike anything in traditional physiology. Among the physiology establishment, still largely dominated by the medical tradition, it has fallen on surprisingly deaf ears. Hill starts from the premise that organisms, and parts of organisms, obey the rules of newtonian mechanics, which is a difficult point to make to those who have forgotten (or never knew) the difference between weight and mass. Decades later, people continue to 'discover' Hill's basic principles by laborious empirical methods. It is not my purpose in this book to review Hill's contributions to science, but rather to take a careful look at his newtonian starting point, and then see whether his method of reasoning can be extended a little bit, to give a window on the dynamics of ecosystems. If such fearless extrapolation leads to errors, that is only to be expected. Finding the reasons for the errors is sometimes more interesting than the original exploration.

I am indebted to Steven Vogel and another anonymous referee for their comments on the first, somewhat confused, draft of this book, which have enabled me to reorganize it, and to eliminate some of its worst short-comings. Following this, a group of students, Sean Kirkpatrick, Sandra Perez, Joe Slowinski, and Catherine Sahley, went through every line and rooted out all manner of errors and misunderstandings, large and small. Finally, my colleague Dave Janos straightened out some of my worst ecological errors. I am deeply grateful to them all. The remaining errors are exclusively my own.

Miami   C. J. P.
1991

Contents

1. Newton still rules

1.1 The scale of biology

Biology covers a well-defined range of scale. The smallest objects of biological interest are large molecules, and the largest is the earth's biosphere. Impressive though this range is, physics covers a wider one. The sizes of subatomic particles are several orders of magnitude below those of the smallest biological molecules, while at the other end of the physical scale, the earth is an indiscernible speck against the background of the universe. In spite of the tremendous range of scales of phenomena that they study, physicists have been far more successful than biologists at integrating the very large with the very small. Astronomers who study distant stars and galaxies do so by observing the radiation which these remote objects emitted at some past time, and then use the theoretical structure of subatomic physics to deduce how the radiation originated. Molecular biologists are fond of reminding those who study biology at larger scales that, likewise, all biology depends ultimately on molecular processes. However, nobody has devised a way to interpret observations of ecosystems directly in molecular terms. At each level of biology, the molecule, the cell, the organ, the organism, the population, and the ecosystem, new principles and sets of rules seem to appear, without any quantifiable connection with the level below.

Newton's once-absolute rule over the physical world has been badly shaken in the last hundred years, at both ends of the scale of size. At the subatomic level, quantum rules prevail, which defy the common sense of us macroscopic creatures, while at cosmic scales, relativity warps the rules of mechanics in equally perplexing ways. However, at the intermediate scales of biology, newtonian physics still works as well as ever it did. The reason is that Isaac Newton was himself a medium-sized animal, and naturally discovered laws that work best over the range of scales which he could perceive directly. The laws of 'modern' physics are concerned with scales that cannot be perceived, and are extremely difficult even to imagine, hence the centuries of effort required to discover them. Biology occupies that range of scale in which newtonian mechanics can account for physical processes, to a level of precision appreciably higher than that to which biologists are accustomed. In biology, if not in physics, Newton still rules.

Modern biology students have to learn chemistry and physics at school, but when they become biologists, most of them use the chemistry and forget the physics. This is unfortunate, as 'classical' physics, of the kind taught in school curricula, has many applications at all the different scales of biology. In the following chapters, I shall consider some processes at various scales, ranging from the intracellular mechanics of muscle fibres to the dynamics of ecosystems. I hope to show first that attention to physical concepts is very helpful in understanding biology at all these levels, and second, that newtonian ideas provide a link allowing processes at each level to be understood in terms of the one below. It is indeed possible to formulate some principles that explain the working of ecosystems in terms of the characteristics of their constituent organisms, although not by considering molecules. These arguments are based on some elementary physical principles, which are all too often neglected by biology students and eminent biologists alike, being perceived (wrongly) as not having any direct application to biological problems.

1.2 Dimensions

I begin with the principle that one should at all times take care to identify the physical character of any variables under discussion. If this appears self-evident to the point of triviality, I beg the reader to suspend judgement for a chapter or two. I shall produce some examples in which careful consideration of physical 'dimensions' can by itself lead to solutions of problems that are otherwise intractable, and several more in which a look at the dimensions points directly to an unfamiliar, but helpful, way of looking at a familiar problem. The basic principle of dimensional reasoning was known by at least the early 1830s, when the engineer Isambard Kingdom Brunel used a simple and crushing dimensional argument at a public meeting, to dispose of a critic who claimed to have proved that his projected steamship, the *Great Western*, would not be able to carry enough coal to cross the Atlantic. Lord Rayleigh, the famous pioneer of aeronautics, is generally credited with having developed the formal technique known as 'dimensional analysis', which became a basic tool in engineering and applied mathematics. Some simple examples of the formal method are presented in Chapter 2, but it is not the purpose of this book to supply a practical guide to the uses and limitations of dimensional analysis. That subject is covered in many engineering textbooks, and McMahon and Bonner (1983) have written a fascinating account of its application to biological problems. My purpose is to apply the basic principle underlying these applications, in a less formal way, to some biological problems.

The notion of a variable's *dimensions* has to be distinguished from that of the variable itself. Newton's Second Law of Motion states that force equals mass times acceleration, or $F = ma$. Here the italicized symbols F, m, and a stand for numbers. m, for example, can be replaced with a number of grams, pounds, etc., to represent the actual mass of some particular object. The dimensions of m are represented by a bold-face capital **M** (for mass), using a roman rather than an italic font, to indicate that this is not a variable that can be replaced by a number. **M** is a statement which identifies the physical character of the variable *m*, but has nothing to do with its magnitude. Similarly, velocity (any velocity) has the dimensions of length (**L**) per unit time (**T**), giving **LT**$^{-1}$. The variable *a* above, which is an acceleration, or rate of change of velocity, therefore has the dimensions of velocity/time, that is **LT**$^{-2}$.

To put this in a more compact notation, we can write Dim (*x*) to mean 'the dimensions of the variable *x*', so that Dim (*m*) = **M**, and Dim (*a*) = **LT**$^{-2}$. In an equation which relates physical variables to each other, the 'equals' sign is really a verb which says that the expression on the left-hand side of the sign is identical with that on the right. This is true of the dimensions, as well as of the magnitudes of the variables. Therefore, if, as Newton says,

$$F = ma, \tag{1.1}$$

for the magnitudes, then it is also true that

$$\text{Dim}\,(F) = \text{Dim}\,(m) \times \text{Dim}\,(a). \tag{1.2}$$

that is, force is a variable with the dimensions:

$$\text{Dim}\,(F) = \textbf{MLT}^{-2}. \tag{1.3}$$

1.3 Units and systems of units

The units in which the magnitude of some variable *x* is measured have to reflect the dimensions of the variable. Time can be measured in seconds, minutes, or weeks, as convenient, and length (or distance) in metres, feet, inches, microns, light-years, or any units that have the dimensions of length. So long as the dimensions are right, the choice of units for a particular purpose is largely a matter of custom. In America, for example, pilots think of altitudes in feet and horizontal distances in nautical miles, while motorists think of speed in statute miles per hour, because the speedometers in their cars are so calibrated. Nearly all Americans think of areas in square feet (for floors) or acres (for land), dieticians think of energy in calories, meteorologists think of temperatures in degrees Celsius but convert them into degrees Fahrenheit for public consumption – and then there is mass. Should mass be measured in pounds, or are pounds the units of weight? What is the difference anyway?

It was many years after Newton's death before the fact that weight and mass are two different variables was formally recognized in a rational system of units. The modern unit system originated in France after the French Revolution. The slow and intermittent progress of its development, and adoption for scientific work (still far from complete), has been described in an erudite history by Klein (1974). Most people, scientists and public alike, are far from pleased by the prospect of changing the measuring habits of a lifetime in favour of some 'standard' set of units. If you have written a famous physiology textbook that refers to energy in calories, and generations of students have learned their trade from it, and your colleagues and the editors of your favourite journals also think in calories, then why should you have to change to unfamiliar joules? If you only want to measure amounts of energy, and use them as a form of physiological currency, then why indeed? The calorie is as good a unit as any other for those who do not need to know that energy is basically the same as work, and work is force times distance. Those who want to calculate amounts of energy (as distinct from just measuring them), do so by multiplying amounts of force by amounts of distance. Such calculations are more convenient if one unit of energy is equal to the product of one unit of force times one unit of distance.

Such simple relationships between units for quantities with different dimensions are the basis of any internally consistent *system* of units. The construction of a system of units for mechanical measurements must begin with the selection of three 'fundamental' quantities, whose units are defined arbitrarily, after which the units for other, 'derived' quantities are made up from combinations of those chosen three. Most people think that there are two such systems of units in common use, the imperial (foot–pound–second) system, and the metric (metre–kilogram–second) system. Actually, there are four, because the metric and the imperial systems each come in two versions, which differ in that radically different meanings (and dimensions) are assigned to the kilogram and the pound. There are *two* unit systems based on the kilogram, metre, and second, only one of which is the SI (Système Internationale). Unfortunately, many of our deeply ingrained habits in 'weighing' objects and so on, are based on the other system, which is not compatible with the SI system, because its 'kilogram' has different dimensions.

1.4 Weight and mass — SI and Engineering unit systems

The SI unit of mass is the kilogram (kg), and that of acceleration is the metre per second-squared ($m\,s^{-2}$). The unit of force is derived by multiplying these two units together, to make the $kg\,m\,s^{-2}$ (corresponding with the dimensions in Equation 1.3). This unit also has a shorter name, the newton (N). One newton is that amount of force which, when applied to a mass of 1 kg, will cause it to accelerate at $1\,m\,s^{-2}$. Typical objects, when dropped near the earth's surface, accelerate downwards initially at about $9 \cdot 81\,m\,s^{-2}$ (with small variations from place to place). From Equation 1.1, it follows that each kilogram of the object is subjected to a gravitational force of $9 \cdot 81$ N. This force is its weight. In the SI system, the unit of weight is the newton. It is wrong to measure weights in kilograms. When biologists 'weigh' objects, they most often want a measure of the amount of matter in the object. This concept is the mass, not the weight, and kilograms (or grams, etc.) are the proper units to use in the SI system. However, when engineers express the 'weight' of an object in kilograms, they usually mean what they say, that is the force which gravity exerts on the object. In the 'engineering' version of the metric system the kilogram is a unit of force. It cannot simultaneously be a unit of mass, and therefore engineers who use this system must have a separate, derived, unit to measure mass.

As noted above, to construct a self-consistent system of units, three quantities have to be selected (arbitrarily) as 'fundamental'. In the SI system the three fundamental quantities were chosen to be mass ($\mathbf{M}$), length ($\mathbf{L}$), and time ($\mathbf{T}$). In this system, force (including weight) is a derived quantity, with dimensions $\mathbf{MLT^{-2}}$. The engineering version of the metric system starts from a different choice of fundamental quantities. Force ($\mathbf{F}$) was chosen to be a fundamental quantity, along with length ($\mathbf{L}$) and time ($\mathbf{T}$). In this case, mass has to be a derived quantity. By turning Equation 1.2 around, you can see that the dimensions of mass in this system are $\mathbf{FL^{-1}T^2}$. In the 'engineering' or 'technical' version of the metric system, the kilogram is a unit of *force*, and the unit of mass is the $kg\,m^{-1}\,s^2$, also known as the 'metric technical mass unit', or 'metric slug'. This is a large unit, equal to about $9\cdot81$ of the SI system's mass-kilograms. In the same way, there are two versions of the old British imperial system, both of which still survive in the USA. The 'physical' version corresponds to the SI system, with the pound arbitrarily defined as the fundamental unit of mass, and a derived unit of force, the poundal, corresponding to the newton. In the 'engineering' version, the pound is arbitrarily defined as the fundamental unit of *force*, and a large derived unit (the slug) is then needed for mass. The dimensions of some commonly used quantities, according to the conventions of the SI system, are given in Table 1.1. Readers who require the dimensions in the engineering system can find a selection in Pennycuick (1988). The origins and histories of the units mentioned, and many others, have been chronicled by Klein (1974).

When it comes to practical definitions of the fundamental units, all four unit systems make use of one or other of two lumps of metal, arbitrarily named the 'standard kilogram' and the 'standard pound'. In the SI system the kilogram is defined as the *mass* of the standard kilogram, while in the engineering metric system the fundamental unit is also called the kilogram, but is defined as the *weight* of this same standard kilogram. The definition of the engineering kilogram (or kilogram-force) as the weight of a particular object requires a 'standard' value to be assumed for gravity, whereas no such assumption is required in the SI system. This is a logical defect in the engineering system, and because of it, the engineering establishment has come to regret the choice of system handed down to them by past generations of engineers. Modern engineering students are taught to do

Table 1.1 Dimensions of some variables used in this book, according to the conventions of the SI System

Variable	Equivalent to	Dimensions
Mechanical variables:		
Mass		M
Length, distance		L
Area		L^2
Volume		L^3
Time		T
Density	Mass/volume	ML^{-3}
Velocity	Distance/time	LT^{-1}
Acceleration	Velocity/time	LT^{-2}
Force	Mass × acceleration	MLT^{-2}
Stress	Force/area	$ML^{-1}T^{-2}$
Strain	Length/length	Dimensionless
Strain rate	Strain/time	T^{-1}
Work	Force × distance	ML^2T^{-2}
Power	Force × speed or work × frequency	ML^2T^{-3}
Volume-specific work	Work/volume	$ML^{-1}T^{-2}$
Mass-specific work	Work/mass	L^2T^{-2}
Volume-specific power	Power/volume	$ML^{-1}T^{-3}$
Mass-specific power	Power/mass	L^2T^{-3}
Ecological variables:		
Biomass		M
Biomass density	Mass/area	ML^{-2}
Production (1)	Mass/time	MT^{-1}
Productivity (1)	Mass/(time × area)	$ML^{-2}T^{-1}$
Production (2)	Energy/time	ML^2T^{-3}
Productivity (2)	Energy/(time × area)	MT^{-3}
Mass inflow	Mass/time	MT^{-1}
Intake rate	Mass inflow/mass	T^{-1}

problems in the SI system, even in America. However, changing the habits of their elders is more difficult. The three non-SI unit systems will remain with us for some time yet, and while they survive, so will the confusion they cause.

1.5 Keeping track of dimensions

It is particularly unfortunate that two entirely different units, with different dimensions, are in common use under the same name (the kilogram), besides a similar pair of units under another name (the pound). Engineers are always explicit as to which unit system they are using, if necessary specifying kilograms-mass or kilograms-force as the case may be, but biologists are not always so conscientious. Often they 'weigh' things, and report the 'weight' in grams, when they actually mean the mass. Ecologists have been known to state that the biomass of a population may be estimated by multiplying the average weight of an individual by the number in the population. The prefix 'bio-' means that the measurement refers to living material, and does not alter the physical dimensions of mass. For some reason nobody talks about 'bioweight'. It is true that there are areas of biology in which no actual errors result from this slipshod habit, but care is needed when weights or masses are combined with other variables with different dimensions. The importance of this is not restricted to such mechanical matters as calculating power requirements for locomotion. The functioning of ecosystems can also be better understood if careful attention is paid to the dimensions of variables. Table 1.1 summarizes the dimensions of variables used later in this book, including some ecological variables used in Chapters 6 and 7.

In the following chapters I shall frequently return to the importance of paying more careful attention to the dimensions of all variables than is customary in biology. Often the key to presenting a problem in a manageable form is to convert the 'raw' variables, in terms of which the original measurements were made, into others with different dimensions. I shall follow this theme through some well-known (and less well-known) topics in the mechanics of muscles and locomotion, through a discussion of the effects of scale, to arrive at a somewhat unconventional way of looking at ecological processes.

2. Gravity, frequency, and the method of dimensions

2.1 Gravity

Gravity can be represented in various ways. The direct way in which we perceive gravity is as the ratio of any object's weight to its mass. The value of this ratio depends on the surroundings. Until recently, people (at least in historical times) had only experienced a narrow range of values for this ratio, around $9\cdot81\,\mathrm{N\,kg^{-1}}$. Since the coming of space-flight, a few privileged explorers have walked in reduced gravity on the surface of the moon, and many astronauts have experienced zero gravity relative to a spacecraft in orbit. Through the miracle of television, we have all been spectators of these unprecedented adventures. No longer do we have to exert our imagination to see that weight and mass are two different things. The astronaut in orbit still has the same mass as he had before he was launched, but is temporarily deprived of his weight.

Many biologists still find it difficult to think of gravity as a variable, because, in their practical experience, gravity is constant. The effects of varying gravity are not accessible to those empirical methods of investigation that are most familiar in biology. The empirical way to investigate the effect of one variable on another is to manipulate the value of the first variable, and observe what happens to the second. For example, if you want to know the effect of, say, carbon dioxide concentration on plant growth, you vary the carbon dioxide concentration, and observe the resulting changes in plant growth. Space flight apart, that method cannot be used to investigate the effects of gravity. Consequently many biologists feel that, if gravity cannot be manipulated, then it cannot be a biologically significant variable. However, our particular value of gravity shapes and constrains in many ways the kinds of organisms that live on the surface of the earth, and it is readily conceivable that life exists in other places where gravity is higher or lower than on earth. There are also reasons to believe that the strength of gravity at the earth's surface has varied by quite large amounts over geological time, with effects that can be seen in certain fossil creatures.

An analytical rather than an empirical approach, as favoured in the physical sciences, can shed some light on the effects of altering gravity, even though the opportunities for experimental verification are limited as yet. Some important effects of gravity can be unambiguously identified, simply by inspecting dimensions.

2.2 Frequency as a fundamental variable

Frequency is another variable which for some reason is used less by biologists than by, for example, electrical engineers. Frequency has the dimensions of inverse time $(\mathbf{T}^{-1})$, and is really a more fundamental variable than time itself. Einstein's famous definition of time (that which you measure with a clock) leads to an obvious follow-up question, which is seldom if ever addressed in physics textbooks: what, exactly, is a clock? A clock consists of two components, an oscillator, and a counter to count the cycles. An oscillator is a device which performs some cyclic process at a well-defined 'natural' frequency, which is assumed to be constant. Nowadays the natural frequency of an atomic clock is the ultimate standard on which measurements of time are based. 'Natural' frequencies abound in living systems as well as in the physical world, and many of them are of the utmost importance in understanding how living organisms work. In studies of animal locomotion, it is often important to know how some natural mechanical frequency is affected by changing the size of the animal, or by changing gravity or some other variable.

2.3 The formal method of dimensions

McMahon and Bonner (1983) have described the formal method of dimensions, with applications to a huge variety of biological problems, and a richly illustrated account of the history of the subject. I shall restrict myself to introducing some simple problems in animal locomotion, involving gravity and frequency among other variables, to show how the classical method works, and to indicate its limitations when faced with the untidiness of biology. This is not intended as a primer in the use of dimensional analysis (which may be found in engineering textbooks), but rather as background for the reasoning to follow. Although I shall seldom invoke dimensional analysis as such in later chapters, the reasoning depends on keeping track of dimensions, to identify the most appropriate variables for representing various problems, which is often half the battle.

The formal method depends on the proposition that both sides of an equation must have the same dimensions. One can sometimes show that, to satisfy this requirement, the variables on the right-hand side of an equation *must* be combined in a particular way, in order to produce dimensions that match those of the variable on the left. To take a very simple example, we may suspect that the frequency at which a pendulum swings is determined by only a few physical variables, one of which is gravity. On closely examining a typical pendulum, it seems likely that only two other variables are involved, the length of the string and the mass of the bob. We then assume that there is an equation relating the frequency—on the left of the 'equals' sign—to the three independent variables, combined in some

Table 2.1   Variables affecting pendulum frequency

Variable	Symbol	Dimensions
Frequency	f	T^{-1}
Length of string	ℓ	L
Mass of bob	m	M
Strength of gravity	g	LT^{-2}

unknown way on the right. The first step is to list the variable that we want to calculate, and the variables that we hope to calculate it from, with algebraic symbols to represent their magnitudes, and dimensional formulae to represent their dimensions (Table 2.1). The frequency of oscillation is the reciprocal of the period (dimensions T), and therefore has dimensions T^{-1}, while the dimensions of gravity (LT^{-2}) are the same as for acceleration (Chapter 1). We assume that the frequency can be found by multiplying the other variables together, having first raised each to some unknown power, thus:

$$f = K\ell^{\alpha}m^{\beta}g^{\gamma}. \tag{2.1}$$

Equation 2.1 says that the magnitude of the frequency is the same as the magnitude of the expression on the right, and also that both sides have the same dimensions. The method of dimensions consists of finding values for α, β, and γ, which make the dimensions of the right-hand side equal to T^{-1}. The number K, which is required to make the magnitudes equal, has no dimensions, and its numerical value cannot be found by this method. The equation relating the dimensions on the two sides does not involve K, but it does involve the three exponents, α, β, and γ:

$$T^{-1} = L^{\alpha}M^{\beta}(LT^{-2})^{\gamma}. \tag{2.2}$$

We have to compare the exponents of M, L and T on the right and left sides of the 'equals' sign. M and L are not represented in the dimensions of frequency (on the left-hand side), but we could express their absence by saying that M and L are both raised to the power zero, thus:

$$M^{0}L^{0}T^{-1} = L^{\alpha}M^{\beta}(LT^{-2})^{\gamma}. \tag{2.2a}$$

This is the same as Equation 2.2, but draws attention to the fact that the exponent of M is zero on the left, and β on the right. The 'equals' sign says that zero is the same as β. We can collect up the exponents on both sides, and write three equations, for the exponents of M, L and T, respectively:

$$\text{Exponents of } M: \quad 0 = \beta; \tag{2.3a}$$
$$\text{Exponents of } L: \quad 0 = \alpha + \gamma; \tag{2.3b}$$
$$\text{Exponents of } T: \quad -1 = -2\gamma. \tag{2.3c}$$

Equation 2.3a shows that contrary to our original suspicions, the mass of the bob is not involved in determining the frequency. The frequency turns out to vary with m^0, that is, it is independent of m. The effect of gravity on the frequency is uniquely determined by Equation 2.3c, because gravity is the only one of the three variables whose dimensions involve time: $\gamma = 1/2$. Substituting this in Equation 2.3b, $\alpha = -1/2$. Thus Equation 2.1 becomes

$$f = Kg^{1/2}\ell^{-1/2} \qquad (2.4)$$

in other words, f is proportional to $\sqrt{(g/\ell)}$.

This argument does not supply a value for the dimensionless multiplier K. There are two ways to find K. We can do a dynamical calculation as in any calculus book (from which it turns out that $K = 1/2$), or we can set up at least one pendulum, measure gravity and the length of the string, and time its oscillations. Probably, most biologists' natural inclination would be to measure the frequency of dozens of pendulums with different lengths and strengths of gravity, and then do a bivariate regression of f on ℓ and g. As far as ℓ is concerned, you could do that, but trying out different strengths of gravity would involve a lot of space travel around the solar system. It hardly seems worth the expense as we already know how f depends on g. We found that out (above) just by looking at the dimensions, without even needing to resort to calculus. It is not necessary to be able to calculate the absolute value of the frequency, in order to make use of the relationship. Geophysicists use it to determine variations of gravity at different points on the earth's surface, which they do by observing the frequency of a pendulum to a high degree of precision, and then assuming that the strength of gravity varies with the square of the frequency. There are many biological examples in which similarly incomplete information can be put to use.

The same method can be directly applied to some problems in biology, for example the stepping frequency of walking mammals, which is important because it determines the power that can be produced by their muscles (Chapter 3). A walking mammal is very much like an inverted pendulum, with the mass above the centre of rotation instead of below it, swinging in an arc centred on the foot. Alexander (1980) has analysed this motion in the full newtonian manner, and has shown, among other things, that the stepping frequency, in animals of similar morphology but different size, varies in proportion to $\sqrt{(g/\ell)}$, where ℓ is the leg length. This result could have been deduced from the same dimensional argument given above for a pendulum. Once again, we could have begun by assuming that the animal's body mass would be involved as well as the leg length, but the argument would have shown that we only have to consider the leg length and gravity, not the body mass.

2.4   Problems with more than three variables

It is an inherent limitation of the method of dimensions that it produces three equations (for the exponents of **M, L,** and **T**), and therefore cannot supply explicit solutions for more than three unknowns. When a problem involves more than three variables, engineers resort to the Buckingham 'pi-theorem', which permits a number of separate variables to be combined into a smaller number of dimensionless ratios, so reducing the number of variables which have to be explicitly considered. There is an especially good account of this method in Anderson (1984), who illustrates it by showing how the force on a body in a moving fluid depends on two such dimensionless ratios, the Reynolds number and the Mach number. However, the reader should not be misled by the elegance and simplicity of that well-known argument. It is true that if a physical variable is assumed to be a function of a number of other physical variables, application of the pi-theorem will always yield the specified number of dimensionless ratios, but it is not always true that the ratios so determined have the profound physical significance of the Reynolds and Mach numbers. In biology, this is often because the variables themselves are not related to one another in a simple and direct way. For example, a bird's wing-beat frequency is a simple physical variable, and it certainly depends on such other easily measurable variables as the air density, gravity, and the wing span of the bird. However, the flapping motion is actually controlled by signals sent from the bird's nervous system. Its frequency, although limited by physical variables, is not determined by them as directly as that of a pendulum. While the pi-theorem will always supply dimensionless ratios which should simplify the problem, in practice they often fail to provide much clarification, if (as usual in biology) the problem has been represented as being physically simpler than it actually is.

At the opposite extreme is the typical biologist's approach, in which one simply measures the wing-beat frequency of every available bird, and runs a multivariate regression against every known variable that could conceivably affect the result, paying no attention whatever to dimensions. I suggest that it is never admissible to neglect dimensions, but also that biological problems often do not satisfy the implied criteria for valid application of strictly physical methods like the pi-theorem. A hybrid approach is indicated. If there are more than three variables, it is not possible to identify the values of the exponents uniquely, as in Section 2.3 above, but it is possible to use the same procedure to find relationships between different exponents, which restrict the range of possible solutions. Empirical methods can then be used to resolve any remaining ambiguities. This approach has two advantages over the use of statistical methods by themselves. First, it allows the inclusion of variables like gravity which resist the empirical approach, and second, it ensures that the final result is dimensionally correct. This latter precaution may be unfamiliar to many biologists.

I stress its importance here, because it is the basis of the reasoning in the rest of this book.

2.5 The wing-beat frequency of birds

The frequency at which birds of different shapes and sizes flap their wings is important for the same reason as the stepping frequency of walking mammals (it determines the power available from the muscles), but more physical variables are involved in this case. The body mass (or rather the weight) does affect the flapping frequency, because a bird has to support its weight in a fluid (air) which is much less dense than its body. A heavier bird must flap harder than a lighter one, other things being equal, in order to support its weight and prevent itself from accelerating earthwards like other, more typical, objects. Other variables that play a part in determining the frequency are the wing span, the wing area, the moment of inertia of each wing about the shoulder joint, and a second environmental variable, the density of the air. We can tabulate these variables as before (Table 2.2). We assume, in the same way as before, that

$$f = K(mg)^\alpha b^\beta S^\gamma I^\delta \rho^\varepsilon, \tag{2.5}$$

from which the equation for the dimensions is

$$\mathbf{T}^{-1} = (\mathbf{MLT}^{-2})^\alpha \, \mathbf{L}^\beta \, (\mathbf{L}^2)^\gamma \, (\mathbf{ML}^2)^\delta (\mathbf{ML}^{-3})^\varepsilon. \tag{2.6}$$

Once again, we can decompose this into three equations for the exponents of $\mathbf{M}$, $\mathbf{L}$, and $\mathbf{T}$:

Exponents of $\mathbf{M}$: $0 = \alpha + \delta + \varepsilon;$ (2.7a)
Exponents of $\mathbf{L}$: $0 = \alpha + \beta + 2\gamma + 2\delta - 3\varepsilon;$ (2.7b)
Exponents of $\mathbf{T}$: $-1 = -2\alpha.$ (2.7c)

The method supplies three equations, as before, but this time we have five unknowns. One of them can be immediately found: $\alpha = 1/2$, from Equation 2.7c. Once again, gravity is the only one of the five variables whose dimensions involve time, and it is also the only one whose effect on the frequency is unambiguously determined. Substituting this value into the other two

Table 2.2   Variables affecting wing-beat frequency

Variable	Symbol	Dimensions
Body weight	mg	$\mathbf{MLT}^{-2}$
Wing span	b	$\mathbf{L}$
Wing area	S	$\mathbf{L}^2$
Wing MI	I	$\mathbf{ML}^2$
Air density	ρ	$\mathbf{ML}^{-3}$

equations, we can say that the remaining four exponents have to be related by the equations

$$\delta + \varepsilon = -1/2; \tag{2.8a}$$

$$\beta + 2\gamma + 2\gamma - 3\varepsilon = -1/2. \tag{2.8b}$$

Although it is true that there are an infinite number of combinations of values of β, γ, δ, and ε which satisfy Equations 2.8, these equations are far from useless. They can be combined with a modest amount of empirical information to yield good estimates for all the exponents. Wing-beat frequencies can be observed in the field for a number of bird species, whose body masses, wing spans, and wing areas are known, and this was done by Pennycuick (1990). The wing moments of inertia unfortunately were not known in this study, but the following stratagem avoided the difficulty. It was assumed that the moment of inertia of a bird's wing is proportional to the body mass, and to the square of the wing span:

$$I \propto mb^2. \tag{2.9}$$

The thinking behind this assumption will come to light later (Chapter 4), and recent measurements by Kirkpatrick (1990) have confirmed that it is not far from the truth. If we use it to substitute for the moment of inertia in Equation 2.5, the moment of inertia no longer appears as such, and the exponents of mass and wing span are now combinations of the original exponents:

$$f = K m^{\alpha+\delta} g^\alpha b^{\beta+2\delta} S^\gamma \rho^\varepsilon. \tag{2.10}$$

From this point, more traditional biological methods can be applied. Suppose that we transform the observed variables into their logarithms, and do a trivariate regression with $\ln(f)$ as the dependent variable, and $\ln(m)$, $\ln(b)$, and $\ln(S)$ as independent variables. From Equation 2.10, the partial regression coefficient of $\ln(f)$ on $\ln(m)$ will be an estimate of $\alpha + \delta$, that of $\ln(f)$ on $\ln(b)$ will be an estimate of $\beta + 2\delta$, while that of $\ln(f)$ on $\ln(S)$ will be an estimate of γ.

The partial regression coefficients, calculated empirically from the field data, are shown in Table 2.3. The 95 per cent fiducial limits show a wider scatter than would occur with simple pendulums, which is due to the fact that

Table 2.3 Partial regression coefficients and their fiducial limits

Variables	Partial coefficient	Fiducial limits	Assumed value
Frequency v. mass ($\alpha + \delta$)	$b_{fm} = 0 \cdot 355$	$0 \cdot 266$ to $0 \cdot 444$	1/3
Frequency v. span ($\beta + 2\delta$)	$b_{fb} = -1 \cdot 32$	$-1 \cdot 732$ to $-0 \cdot 908$	-1
Frequency v. area (γ)	$b_{fs} = -0 \cdot 0886$	$-0 \cdot 310$ to $0 \cdot 133$	$-1/4$

the variables are not related to one another in such a direct, physical way as the method implicitly assumes. The coefficients give estimates of various combinations of the exponents as above. There are now two constraints on possible values of the exponents:

(1) only combinations that satisfy Equations 2.8 will produce a result with the correct physical dimensions;

(2) only combinations that fall inside the fiducial limits shown in Table 2.3 are consistent with the field observations.

There is a limited range of values that will satisfy both requirements simultaneously, including this combination:

$$f = K(mg)^{1/2} b^{-2/3} S^{-1/4} I^{-1/6} \rho^{-1/3}. \tag{2.11}$$

If we approximate for the wing moment of inertia in terms of body mass and wing span, as in Proportionality 2.9, we get this version of the same equation:

$$f = K_1 m^{1/3} g^{1/2} b^{-1} S^{-1/4} \rho^{-1/3}. \tag{2.12}$$

The dimensionless constant may be different from before, and has been changed to K_1. As a small amount of trial and error will show, there is not very much scope for varying the values of the exponents. α is fixed by Equation 2.7c. The remaining four exponents cannot be changed independently of each other. If a different value is tried for any one of them, then the others have to be adjusted so that Equations 2.8 are still satisfied, that is, so that the right-hand side of Equation 2.12 still has the dimensions T^{-1}. After only a small amount of change, one or other of the combinations of exponents in Table 2.3 strays outside the fiducial limits set by the regression calculation. The regression coefficients are only estimates of certain combinations of exponents. The calculated values represent the most probable values, but a range of variation is not excluded, as indicated by the fiducial limits.

While this calculation lacks the elegance of classical applications of dimensional analysis, it also avoids the more absurd results that follow from the uncritical use of regression methods. A straightforward regression procedure could have been applied to those independent variables which can be measured and varied, that is to the body mass, wing span, and wing area, but not to gravity or air density. A regression equation would have resulted, like Equation 2.12 but relating the wing-beat frequency only to the three variables m, b, and S. The left-hand side of this equation would have had dimensions T^{-1} for frequency, as before, while the dimensions of the right-hand side would be $M^{0.355} L^{-1.50}$. This makes no sense from a physical point of view. The order of precedence is to satisfy the dimensional requirement first, and then to adjust the result to accommodate the empirical observations, which is how Equations 2.11 and 2.12 were found.

2.6 Animals whose weight is supported by water

In the previous examples, I have glossed over what is perhaps the trickiest part of dimensional analysis. How do you decide which variables to include in the analysis? Some engineering texts have a formal mathematical answer to that question, but the practical answer is intuition, guided by a sound understanding of the underlying physical process. You have to make a judgement as to which variables can be safely neglected, because they do not appreciably affect the result, and which must be included, because they do. For example, the forces exerted by the muscles of an animal that walks or flies, supporting its weight in air, are clearly affected by gravity, and so it is natural to assume that the oscillation frequency of its legs or wings will be affected by gravity too. It was this assumption that supplied a variable whose dimensions involve time, without which solutions for the frequencies could not have been found. What about an animal whose weight is supported by immersion in water? A swimming animal, far below the surface, only has to exert forces caused by the relative motion between it and the water. These forces do not depend on gravity. A sudden increase in the strength of gravity is actually quite difficult for a submerged animal to detect, unless it has some compressible organ such as a swim-bladder which responds to the change in hydrostatic pressure. How, then, is the frequency with which a fish beats its tail determined? In walking and flying animals, gravity is the variable that supplies the power of T in the dimensions, but if a swimming animal cannot detect gravity, there is no justification for including it in the calculation. In that case, some other variable, whose dimensions involve time, must be involved.

If we imagine the fish beating its tail more and more rapidly, the *stress* that its muscles have to exert will increase with increasing frequency, by an amount that will itself be greater in a dense fluid than in a less dense one. The stress is the tension exerted per unit cross-sectional area of muscle. Reasons will come to light in Chapter 3 why a series of similar animals of different size, all 'cruising' in a similar manner, might find it necessary to adjust their motion so that the stress exerted by their muscles is the same in all of them. We can identify muscle stress as a variable, whose value has to be held in a certain range, and which helps to determine the tail-beat frequency. In addition to muscle stress, we may provisionally identify body

Table 2.4 Variables affecting tail-beat frequency

Variable	Symbol	Dimensions
Muscle stress	σ	$\mathbf{ML^{-1}T^{-2}}$
Water density	ρ	$\mathbf{ML^{-3}}$
Body length	ℓ	$\mathbf{L}$

length and water density as relevant variables, and list them with their dimensions (Table 2.4). As before, we postulate that the frequency (f) is proportional to the product of these variables, each raised to an unknown power:

$$f = K\sigma^\alpha \rho^\beta \ell^\gamma. \tag{2.13}$$

The equation for the dimensions is:

$$\mathbf{T}^{-1} = (\mathbf{M}\mathbf{L}^{-1}\mathbf{T}^{-2})^\alpha (\mathbf{M}\mathbf{L}^{-3})^\beta \mathbf{L}^\gamma. \tag{2.14}$$

This time we have three equations for three unknowns:

Exponents of **M:** $0 = \alpha + \beta$; (2.15a)
Exponents of **L:** $0 = -\alpha - 3\beta + \gamma$; (2.15b)
Exponents of **T:** $-1 = -2\alpha$, (2.15c)

whence, $\alpha = 1/2$, $\beta = -1/2$, and $\gamma = -1$. The only one of the three variables which is readily accessible to empirical observation is the body length. If it is true that sharks in slow cruising locomotion all exert much the same stress in their muscles, and if the density of sea-water is regarded as constant, then species of different size, from small dogfish to great white sharks, should beat their tails at frequencies inversely proportional to their body lengths. This is not the same rule that applies to walking and flying animals, and the consequences of this difference, assuming the dimensional deduction to be correct, are followed up in Chapters 4 and 5. The ramifications extend not only to speed in locomotion, but also to the geometry of the respiratory organs.

The prediction that tail-beat frequency varies inversely with body length could be checked from video recordings of cruising sharks, taken from a light plane off places like Miami Beach, if some way could be found to determine their lengths. It would be somewhat more difficult to verify the prediction that shark-like animals on planets with gravity higher or lower than that of earth should not show any difference of tail-beat frequency from our own sharks, unless their muscles run at higher or lower stress. This rule also differs from the one for walking and flying animals, whose stepping and flapping frequencies should vary with $\sqrt{g}$ (above). These predictions might seem arcane, but they too have ramifications to which I shall return in later chapters. Their relevance is not confined to the study of extraterrestrial faunas. They are also of interest in interpreting the mechanics of extinct animals, which lived at times when the earth's surface gravity may have been stronger or weaker than it is now.

2.7 Gravity and work

Another way to think about gravity is as a measure of the work that has to be done to lift a mass upwards, i.e. away from the centre of the earth. When one carries a load upstairs, one has to do work at the rate of 9.81 J kg^{-1} m^{-1}. To start a pendulum swinging, work has to be done to pull it to one end of the swing, since the bob has to be raised against gravity. This work is said to be stored in the form of 'potential energy', because it can be recovered by allowing the bob to fall. When the bob is released, it descends and accelerates, converting its initial potential energy into kinetic energy. As it passes the mid-point, it begins to rise, and the kinetic energy is converted back again into potential energy. Neglecting friction, the bob stops (zero kinetic energy) when it has risen back to its original height, and the cycle then repeats indefinitely. If gravity is made stronger, more work has to be done to raise the bob to the same height as before. The bob therefore has more kinetic energy at the mid-point of the swing, that is, it moves faster, and oscillates at a higher frequency. Likewise, a walking animal steps more quickly in stronger gravity. The Apollo astronauts found that their steps were inconveniently slow in the moon's weak gravity, and resorted to hopping like kangaroos in order to move around.

In walking, not all of the kinetic energy at the mid-point of each step is reconverted into potential energy. Some work has to be done to maintain the vertical motions, and the amount of work required at each step increases as gravity becomes stronger. The stepping frequency also increases in stronger gravity, so that these larger amounts of work have to be done more often. Thus the strength of gravity has a strong effect on the *power* required for walking, that is, the rate at which work has to be done. Similarly, a flying animal supports its weight by pushing air away from it downwards. If its weight increases because of an increase in the strength of gravity, it has to push air downwards at a greater rate, and/or accelerate it to a higher downward velocity, and so more power is required to fly. On the other hand, the power required by a submerged animal to swim depends on its speed and the density of the water, but not on gravity.

2.8   Gravity past and present

Biologists are not the only ones who perceive the strength of gravity as a constant. Most geologists also believe that the strength of gravity at the earth's surface has changed by only minor amounts, if at all, over geological time. An exception is Carey (1976), who has assembled a mass of evidence that the earth's radius is much larger now than it was in early Mesozoic times, probably due mainly to a reduction in the density of the minerals making up the mantle. Carey cautions against too readily assuming that any of the earth's physical characteristics have remained constant over long periods, including its mass, radius, plane of rotation, and surface gravity. It will emerge in Chapter 4 that the current value of gravity places some limits on the possible characteristics of animals using different types of locomotion. This is an argument which can be reversed, to make deductions about the past value of gravity from the characteristics of fossil animals. It would seem that gravity has not been constant throughout geological time, although the inferred variations do not exactly coincide with Carey's predictions.

3. Muscles as engines

3.1 The limits of muscle power are mechanical

Animals, apart from very small ones, rely on muscles to propel themselves, and to procure and handle food, among other essential purposes. Muscles are biological engines which consume chemical energy and perform mechanical work. In most physiology textbooks the function of muscles is considered under the general heading of 'metabolism', along with other processes, like thermoregulation, which also consume oxygen and generate heat. Locomotion is seen as a kind of 'metabolism', because animals get hot and consume more oxygen when they run about. The power output of muscles, that is, the rate at which they do work, is often seen as being limited by the rates at which enzyme systems can supply energy. This is putting the cart before the horse, however. The rate at which a muscle can do work is limited by three variables and three only, the stress which it can exert, the strain through which it can shorten, and the contraction frequency. These are mechanical variables, and their maximum values are set by mechanical limitations. Adaptation of enzyme systems and so on is a secondary matter. The rate at which they supply energy has to be adjusted to match the rate at which the mechanical system can absorb it.

3.2 Force and movement produced by a muscle

All muscles can produce force, and some are specialized to do this, as their primary function. The force is always in the form of tension. Muscles can only pull. Unlike many man-made engines, they cannot push, or twist. Muscles shorten actively, but have to be lengthened passively, either by the contraction of another 'antagonist' muscle, or by some elastic structure. Their internal organization is illustrated in Fig. 3.1, and the numbers which follow are based on the account in White and Thorson (1975). At the subcellular level, the contractile material of the muscle fibre is organized as a series of *sarcomeres* which are connected end to end. The sarcomere is the basic contractile unit, and consists of a parallel array of myosin fibres in the middle, interdigitating with two arrays of thinner actin fibres at either end. The actin fibres are connected to the Z-membrane which forms the end of the sarcomere. They pull against the actin fibres of the next sarcomere, which are attached to the other side of the same membrane. The tension is produced by lateral projections (cross-bridges) which project from the myosin strands, and connect to, and pull on the interdigitating actin strands. It seems that each cross-bridge can exert a maximum force of about 5·3 pN.

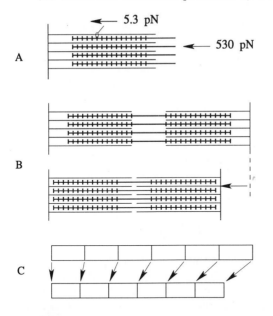

Fig 3.1 A, In round numbers, each cross-bridge can exert a pull of 5·3 pN, and there are about 100 of them on each end of each myosin filament, producing a tension of 530 pN at the middle of the filament. B, The geometry of the sliding filaments permits the sarcomere to shorten through a limited fraction of its extended length (strain), before the myosin filaments butt against the Z-membranes, and the actin filaments butt against each other. C, When many sarcomeres are connected end to end, the total distance shortened is the sum of the contributions from each sarcomere (as is the total speed), but the strain is the same for the whole muscle as for each individual sarcomere.

Typically, in vertebrate striated muscle, each end of each myosin filament would have about 100 cross-bridges pulling in the same direction, so that their individual contributions build up to a tension of about 530 pN at the centre of the filament. This is balanced by the pull of an equal number of cross-bridges at the other end of the myosin fibre, pulling in the opposite direction, on a separate array of actin filaments. The force could be increased by lengthening the myosin filament (so increasing the number of cross-bridges pulling together), but in that case the filament would have to be thicker to carry the additional force, which would reduce the packing density of the filaments, and disturb the cross-sectional geometry of the array. Huxley (1985) states that the myosin filaments are the same length (1·55 μm) in all vertebrate striated muscles, although longer (and thicker) myosin filaments, with a different cross-sectional arrangement are found in some crustaceans.

The myosin cross-bridges continually attach and detach from sites on the actin filaments. The relative rates of attachment and detachment determine the average number of cross-bridges attached at any one time, and hence the force exerted. If the muscle is permitted by its external attachments to shorten, the myosin filaments crawl like tiny centipedes along the actin filaments. The higher the attachment and detachment rates of the cross-bridges, the faster the filaments can crawl, if unrestrained by an external force. The two Z-membranes defining the ends of the sarcomere approach one another at twice the relative speed between the actin and myosin filaments. A complete myofibril consists of a series of sarcomeres connected end to end. If one end is fixed, the distance moved (and the velocity) at the other end is the sum of the contributions of the individual sarcomeres (Fig. 3.1C).

3.3 Choosing variables to be independent of muscle geometry

Although it is simplest to discuss the action of muscles in terms of force and changes of length, these variables have the inconvenient characteristic that they depend on the size of the muscle. A large muscle exerts more force, and shortens through a greater distance than a smaller one of similar shape. The key to analysing the properties of the tissue 'muscle', as opposed to describing the action of one particular muscle, is to choose related variables with different dimensions. Instead of discussing the force which a muscle can exert, we can use the *stress*, which is the force exerted per unit area of the muscle's cross-section. The stress is related directly to the ultrastructure of a particular type of muscle, and, unlike force, does not depend on the size of the individual muscle. Each myosin filament exerts a force of about 530 pN when the muscle is stimulated isometrically, and there are about $5 \cdot 7 \times 10^{14}$ of them per square metre of cross-sectional area. That makes the maximum isometric stress about 300 kPa. The dimensions of stress (force per unit area) are $\mathbf{ML^{-1}T^{-2}}$, and the SI unit is the pascal (Pa), which is the same as the newton per square metre.

Each sarcomere exerts the same *stress* as those connected to each of its ends, so the stress is the same at any point along the fibre. Each sarcomere can shorten through the same distance, but since the sarcomeres are connected end to end, their individual movements add together (Fig. 3.1C). If one end of the muscle is fixed, then the distance through which the other end can move when the muscle contracts is proportional to the number of sarcomeres. Instead of discussing the change of length as such, which depends on the length of the muscle, we choose a different variable, the *strain*, which does not. The strain is the ratio of the shortening distance to the extended length of the muscle, and it has the same value for the muscle as a whole as for each individual sarcomere. Like stress, it is a property of the tissue, and does not depend on the size or shape of the particular muscle. Typical vertebrate locomotor muscles are permitted by skeletal constraints to shorten through about 25 per cent (maximum) of their extended lengths, that is, they can produce a strain up to about $0 \cdot 25$ although they only actually shorten through this amount of strain in maximal exertion, not in sustained, cruising locomotion. Unlike distance shortened (dimensions **L**), strain is dimensionless. It is the ratio of two lengths, and therefore just a number. No units are needed to measure it.

3.4 Volume-specific work

What if we multiply the stress which a muscle exerts by the strain through which it shortens? Stress is force divided by cross-sectional area, while strain is shortening distance divided by initial length. The product of the two is therefore (force times shortening distance) divided by (cross-sectional area times initial length), that is, the work done in a contraction, divided by the volume of the muscle. More succinctly, if a muscle exerts a stress, σ, and shortens through a strain, ε, then the work done by unit volume of the muscle, or *volume-specific work* (Q_v) is:

$$Q_v = \sigma\varepsilon. \tag{3.1}$$

Often it is more convenient to think about the *mass-specific work* (Q_m), that is the work done by unit mass (rather than unit volume) of the muscle. This can be found by dividing the volume-specific work by the density (ρ) of the muscle.

$$Q_m = \sigma\varepsilon/\rho. \tag{3.2}$$

The density of muscle is typically about $1060 \, \text{kg m}^{-3}$.

3.5 Dynamics of the single contraction

If one end of a muscle is fixed when it contracts, the other end moves with a certain speed which of course depends, among other things, on the length of the muscle. For any particular muscle, there is a definite relationship between the force (F) which the muscle exerts when stimulated, and the velocity (V) at which it shortens. This relationship came to be known as 'Hill's equation', from a famous study by A. V. Hill (1938):

$$V = b\,(F_0 - F)/(F + a), \tag{3.3a}$$

where F_0 is the maximum (isometric) tension, and a and b are constants with the dimensions of force and velocity respectively. A noteworthy feature of Hill's (1938) classic paper is the meticulous way in which he identifies the dimensions of all his variables. Unfortunately, this unusual characteristic also sets Hill's work apart from most other physiological literature, both earlier and more recent.

The velocity of shortening (dimensions $\mathbf{LT^{-1}}$) can be divided by the extended length of the muscle to convert it to a *strain rate*, a related variable which has the advantage that it does not depend on the length or geometry of the muscle. Some people think of strain rate as the speed of shortening, expressed in muscle lengths per second. Since strain rate is the ratio of a speed ($\mathbf{LT^{-1}}$) to a length ($\mathbf{L}$), its dimensions are $\mathbf{T^{-1}}$, the same as frequency. The units are actually not 'lengths per second', but just 'per second', or s^{-1}. Strain rate is a pure rate. Hill's equation can easily be converted to express strain rate (ψ) in terms of stress (σ), and in fact this version of Hill's equation has exactly the same form as the original version:

$$\psi = \beta(\sigma_0 - \sigma)/(\sigma + \alpha), \tag{3.3b}$$

where σ_0 is the maximum (isometric) stress. In this version, α is a 'stress constant', equal to the original force constant (a) divided by the cross-sectional area of the muscle, and β is a 'strain rate constant' equal to the original velocity constant (b) divided by the extended length of the muscle.

Hill's equation is basically an empirical description of the behaviour of a muscle under somewhat artificial conditions. The muscle is maximally stimulated, and shortens against an apparatus which maintains the force constant, regardless of the speed. Fung (1981) has expressed some reservations about the validity of applying Hill's equation under the conditions of natural locomotion. However, the arguments which I shall base on Hill's equation in this chapter do not depend on very precise predictions of stress or strain rate, and should not be affected by these doubts.

3.6 Intrinsic speed and maximum power

If the muscle is allowed to shorten without restraint (that is $\sigma = 0$) then the strain rate reaches its maximum possible value, which is given by Equation 3.3b as:

$$\psi_0 = \beta\sigma_0/\alpha. \tag{3.4}$$

This maximum strain rate is a very important property of any particular muscle. It was called the *intrinsic speed* by Hill, and I shall retain this time-honoured term, although it should be remembered that the dimensions of ψ_0 are not those of speed. It is the property that distinguishes one muscle as 'faster' or 'slower' than another. It can be traced to the rate constants that govern the frequency with which myosin cross-bridges attach and detach themselves at binding sites on the actin filaments. The top section of Fig. 3.2 shows graphs of strain rate versus stress, from the modified Hill's equation (Equation 3.3b), for two muscles, one 'faster' than the other. The intrinsic speed for each muscle is the point where the curve meets the y-axis of the graph.

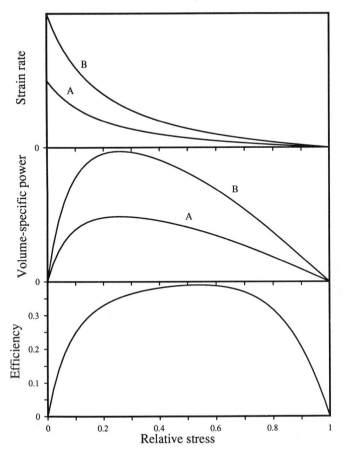

Fig 3.2 Top, A 'fast' muscle (B) and a 'slow' one (A) both produce the same isometric stress (right-hand end of curve), but differ in their maximum strain rates, when the stress is zero. Middle, The same two curves as in the top diagram, converted to volume-specific power by multiplying the stress by the strain at each point on the curve. Bottom, Efficiency v. stress, meaning the fraction of the energy liberated from ATP which is converted into work.

These same curves are replotted in another form in the middle section of Fig. 3.2. The ordinate for each point in the top graph (strain rate) has been multiplied by its own abscissa (stress) to give the new ordinate of the middle graph. As noted above, stress times strain is work done per unit volume of muscle. Stress times strain rate is the volume-spacific *rate* of doing work, that is the *volume-specific power*. At the left-hand end of the graph the muscle is shortening as fast as it is able, but the power is zero, because the stress is zero. At the right-hand end, the muscle is exerting its maximum stress, but the power is again zero, because the strain rate is zero. In between, the muscle produces power. The maximum power is produced when the stress is about 26 per cent of the maximum.

3.7 Efficiency

A muscle is an energy converter. It accepts energy in the form of Gibbs free energy, and converts it into mechanical work on a macroscopic scale. The *efficiency* is measured on a scale from 0 to 1, and is defined as the proportion of the input energy which is converted into work. The efficiency cannot exceed 1, and in practice is considerably less. Some care is needed in defining the 'system' to which any particular discussion of efficiency refers, as there are two distinct stages of energy conversion that take place within a muscle fibre. First, some 'fuel' substrate, usually glycogen or fat, is oxidized, with the end result that a high-energy phosphate group is added to a molecule of adenosine diphosphate (ADP), converting it into adenosine triphosphate (ATP). The second stage takes place within the contractile proteins themselves, where the ATP is reconverted to ADP, and some of the energy so released is converted into work. This second stage is cyclic. A certain amount of chemical energy is consumed in phosphorylating an ADP molecule to ATP, and the original status quo is restored when the molecule is dephosphorylated once again, back to ADP. In this cycle, all of the energy released by dephosphorylating ADP is converted either into work or into heat. If a fraction η of this energy is converted into work, then the rest of it $(1 - \eta)$ is converted into heat. This is approximately true of the first stage also, provided that the energy for rephosphorylating ADP comes from aerobic metabolism of the substrate. In the case of anaerobic metabolism, a substantial amount of the input energy is absorbed in reducing the entropy of the reaction products as compared to that of the reactants, and in that case it is no longer true that the input energy is partitioned into heat and work only. This intricate subject has been reviewed and clearly explained by Gnaiger (1989).

The efficiency of the second (contractile) stage can be reconstructed in considerable detail from Huxley's (1957) quantitative model of the sliding-filament mechanism of muscle, which has itself been reviewed and related to other aspects of muscle function by McMahon (1984). The efficiency depends on the stress and strain. At both ends of the curves of Fig. 3.2 (middle graph) the muscle is doing no work, but is consuming chemical energy. Its mechanical power output, and therefore its efficiency, is zero at these two points, that is, when the stress is either zero, or equal to σ_0. At values of the stress between these limits, the muscle produces work, and the efficiency is positive. The curve of efficiency v. stress (Fig. 3.2, bottom) shows a broad peak at a maximum value of about 0·39, when the stress is about 54 per cent of σ_0. This value for the peak efficiency refers only to

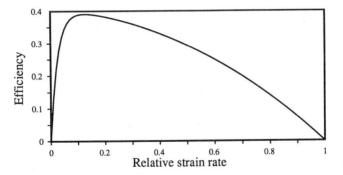

Fig 3.3 Efficiency peaks when the strain rate is about 13 per cent of the maximum for that particular muscle.

the conversion of energy from dephosphorylating ATP into work. Lower values, usually between 0·20 and 0·25, are found in physiological experiments which measure the overall efficiency of converting fuel energy into work.

It is also possible to plot curves of efficiency and power, using strain rate instead of stress as the independent variable (Figs 3.3 and 3.6). The efficiency curve (Fig. 3.3) shows a sharp peak at rather a low value of the strain rate (about 13 per cent of the maximum). If a muscle is forced to shorten at still lower strain rates it becomes inefficient, because it is producing work only slowly, while it still consumes chemical energy. When the strain rate approaches zero, the muscle is effectively maintaining tension rather than producing work, and the concept of efficiency then becomes irrelevant.

3.8 Cost of maintaining tension

Some muscles, known as *tonic* muscles, are adapted to maintain a steady tension without shortening. A muscle doing this consumes fuel energy, but does no work, so its efficiency is zero. This unhelpful observation indicates that efficiency is not the right variable to describe the degree of success with which this type of muscle performs its function. A tonic muscle exerts a force, at the cost of consuming chemical energy at a certain rate. The ratio of this 'power input' (dimensions $ML^2 T^{-3}$) to the force (dimensions MLT^{-2}) is a simple measure of the energetic cost of maintaining force. The ratio of power to force has the dimensions LT^{-1}, i.e. those of velocity. To present the argument in a form which does not depend on the size of the muscle, we can compare volume-specific power (dimensions $ML^{-1} T^{-3}$) with stress (dimensions $ML^{-1} T^{-2}$). The ratio of volume-specific power to stress is a measure of the cost for muscles of a particular type to maintain stress. This ratio has the dimensions T^{-1}. It is a pure rate, like strain rate, and like the rate constants that govern the attachment and detachment frequencies of myosin cross-bridges.

Physiologists have known for many years that 'slow' muscles, i.e. those characterized by a low value for the intrinsic speed (ψ_0), maintain force more cheaply than 'fast' muscles (Johnston 1985). ψ_0 has the same dimensions ($\mathbf{T}^{-1}$) as the energetic cost of maintaining stress (above), and it is tempting to surmise that these two quantities are related. This is indeed the case. McMahon (1984) gives equations from the Huxley sliding-filament model, for calculating the isometric stress, and the volume-specific power needed to maintain that stress. These somewhat intricate equations were compared by Pennycuick (1992), and the outcome was that, when the volume-specific power was divided by the stress, the ratio of the two turned out to be equal to $\psi_0/16$. This implies that there is no need to measure oxygen consumption or heat production, or indeed to make any direct measurement of energy, in order to determine the energetic cost for which a particular muscle can maintain a specified stress. Simple mechanical measurements on a whole muscle will suffice. For example, to determine the energetic cost for which your own biceps can maintain tension, find the maximum speed (in m s^{-1}) at which you can make it shorten, and divide this by the extended length (in metres) to get ψ_0. Divide by 16, and you have the rate of energy consumption per cubic metre of muscle, per pascal of stress maintained. The result is already in units of W Pa^{-1} m^{-3} (actually the same as s^{-1}). This rate of energy consumption refers to the rate at which energy is liberated by dephosphorylating ATP. The rate at which the muscle consumes fuel energy is higher, depending on the efficiency of the oxidative stage of energy conversion.

3.9 Repetitive contraction and work loops

The power output plotted in Figs 3.2 and 3.6 is the rate at which the muscle does work, while it is actually shortening. In sustained locomotion, muscles contract cyclically. Locomotor muscles are almost always arranged in antagonistic pairs. One member of the pair shortens, passively extending its antagonist, then the second muscle shortens, and the first is passively extended. When trying to estimate the rate at which a muscle requires fuel energy, we need its average rate of doing work, which is approximately half the instantaneous rate plotted in Figs 3.2 and 3.6. It would be exactly half if the muscle were to spend half of each cycle shortening, and the other half being passively extended.

To look at it in a slightly different way, the power output can be found by multiplying the work done in one complete cycle of contraction and relaxation (the *cycle work*) by the contraction frequency. The cycle work can be displayed graphically on an oscilloscope screen. In a classic series of experiments, Boettiger (1957) attached one end of an insect flight muscle to a force transducer, while allowing the other end to move against a load, and monitoring its changes of length with another transducer. By suitably choosing the properties of the external load, the muscle was set into oscillatory contraction. Instead of recording the force and movement against time, Boettiger connected the output of the motion transducer to the x-axis of his oscilloscope, and that of the force transducer to the y-axis, so that the moving point on the oscilloscope screen traced out a graph of the force exerted by the muscle, plotted against the muscle's length. Starting at the point A in Fig. 3.4, the muscle is extended (length high), and is developing near-maximum force (tension high). It shortens (length decreasing) while exerting a high force, so doing work on the apparatus. As the length approaches its minimum, the force decreases, then at point B, with the force low, the muscle is lengthened by the apparatus pulling on it. In returning to point A, the apparatus does work on the muscle, but not so much work as the muscle did on the apparatus when it shortened, because the change of length is the same, but the force is less. A simple application of integral calculus shows that the net amount of work done by the muscle on the apparatus, in one complete cycle, is equal to the area of the loop. This might have been suspected from the dimensions. The graph represents length on the x-axis, and force on the y-axis, so any area on the graph must have the dimensions of length times force, that is, work.

The mechanical *power* produced by a muscle performing the same work loop over and over again, as in Fig. 3.4, is found by multiplying the cycle work (area of the loop) by the contraction frequency. As usual the argument

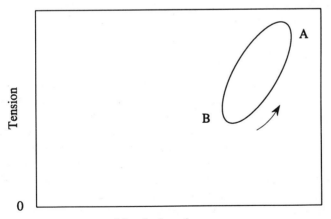

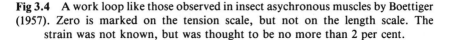

Fig 3.4 A work loop like those observed in insect asychronous muscles by Boettiger (1957). Zero is marked on the tension scale, but not on the length scale. The strain was not known, but was thought to be no more than 2 per cent.

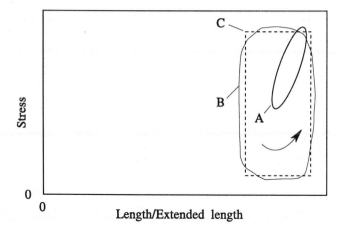

Fig 3.5 A, the same curve as Figure 3.4, converted to stress versus strain, with the zeros marked on both scales. B, Vertebrate locomotor muscles are thought to have less residual stress during lengthening than insect flight muscles, and also to shorten through a greater strain, perhaps 15 per cent in cruising locomotion, and 25 per cent maximum, thus producing more volume-specific work (bigger area of the loop). C, Representing the volume-specific work as a stress-strain product amounts to replacing the real work loop with a rectangular one of the same area.

can be made more general by a suitable choice of variables. If stress is plotted against strain instead of force against length (Fig. 3.5), the diagram looks the same as before, but now the area of the loop represents volume-specific work. Multiplying this by the contraction frequency gives the volume-specific power, which can be converted into mass-specific power by dividing by the density of muscle (about 1060 kg m^{-3}). Work loops like that of Fig. 3.5A have been observed in preparations of isolated insect flight muscles. This work loop is small because the strain is small (typically 0·02), and there is a substantial amount of residual stress while the muscle is being forcibly lengthened. A work loop like this means that each gram of muscle does only a small amount of work in each cycle, but in spite of that, the specific power output of insect flight muscles can be very high, because they operate at high contraction frequencies. No such direct observations have been made on vertebrate locomotor muscles, but indirect evidence suggests that the work loop in bird flight muscle in cruising flight may be like that of Fig. 3.5B, with the stress in shortening being about 150 kPa more than that in lengthening, and the strain about 0·15. The simplified work loop of Fig. 3.5C represents the same cycle work, without attempting to give an account of the detailed (but unknown) time course of stress and strain in the real muscle. If we think of the work loop as rectangular as in Fig. 3.5C, found by multiplying a stress (σ) by a strain (ε), then the volume-specific power output (P_v) if the muscle repeats the loop at a frequency f is

$$P_v = \sigma\varepsilon f, \tag{3.5}$$

and if the density of the muscle is ρ, then the mass-specific power output (P_m) is

$$P_m = \sigma\varepsilon f/\rho. \tag{3.6}$$

3.10 Matching the muscle to the load

Equation 3.3b can be rearranged to give the stress that the muscle will develop while contracting, if it is allowed to shorten at different strain rates. The idea is that the external load imposes some value ψ of the strain rate, and Hill's equation then determines what stress the muscle can develop. Rearranging Equation 3.3b:

$$\sigma = (\beta\sigma_0 - \alpha\psi)/(\psi + \beta). \tag{3.7}$$

The volume-specific power output while the muscle is actually shortening is $\sigma\psi$. In repetitive contraction, the muscle is shortening about half the time, so, as a rough approximation, $P_v = \sigma\psi/2$, or

$$P_v = \psi(\beta\sigma_0 - \alpha\psi)/2(\psi + \beta). \tag{3.8}$$

ψ can take only a limited range of possible values, from zero (at which the muscle develops the isometric stress σ_0) up to ψ_0 (the intrinsic speed) when the stress is zero. We can replace ψ by a dimensionless variable $q = \psi/\psi_0$, which always varies from 0 to 1, for any muscle, fast or slow. Re-expressing Equation 3.8 in terms of q (which may be called the 'relative strain rate') the volume-specific power is:

$$P_v = \sigma_0\psi_0(q - q^2)/2(Kq + 1). \tag{3.9}$$

The constant K is the ratio σ_0/α which typically has a value of about 5 in vertebrate skeletal muscles. Tinkering with this value has very little effect on the curves of P_v v. q. Fig. 3.6 shows such curves for four different muscles, all having the same value of σ_0, but different values of ψ_0, in the ratio 1:2:3:4.

To understand Fig. 3.6, you have to distinguish carefully between the *relative* strain rate (q), which varies from 0 to 1 for all four muscles, and the *absolute* strain rate, which is also zero at the left-hand end for all the muscles, but takes different values elsewhere. The maximum (absolute) strain rate (or intrinsic speed) for each muscle corresponds with $q = 1$, and gets progressively higher as you go from the slowest to the fastest muscle, in the ratio 1:2:3:4. For any value of q between 0 and 1, the absolute values

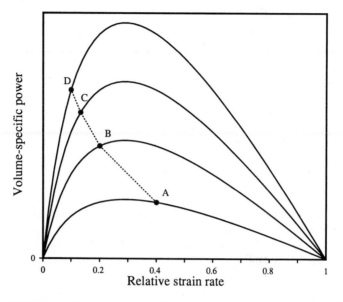

Fig 3.6 Volume-specific power v. strain rate for four muscles, whose intrinsic speeds (ψ_0) are in the ratio 1:2:3:4. The points A–D represent a constant *absolute* value of the strain rate—the faster the muscle, the slower the *relative* strain rate to which this corresponds. Muscle D is too fast to produce power efficiently at this strain rate (compare Figure 3.3). A still faster muscle would produce a little more power, but at the expense of a drastic increase in the rate of fuel consumption.

of the strain rate for the four muscles have the same ratios to each other, and so also do the corresponding values of the volume-specific power, plotted on the *y*-axis. This is because the *stress* is the same for all four muscles for a given value of *q*, although the absolute strain rate is four times higher in the fastest muscle than in the slowest. Since the volume-specific power (instantaneous) is found by multiplying the stress by the strain rate, it too is four times higher in the fastest muscle than in the slowest, at the same value of the relative strain rate.

Fig. 3.6 shows what would happen if you were trying to design an animal, whose muscles are obliged to shorten at some given value of the *absolute* strain rate, because of mechanical constraints, over which the animal has no control. The diagram supposes that you try out four muscles with different intrinsic speeds, and see how much volume-specific power you can get out of each, when it shortens at a fixed value of the absolute strain rate. At first sight it appears that the power output of each muscle would be proportional to its intrinsic speed, but this is not necessarily so. The difficulty is that an absolute strain rate which is 40 per cent (say) of the slowest muscle's intrinsic speed (point A on the lowest curve) is only 20 per cent of the maximum for the muscle of the second curve (point B). The absolute strain rate in the same for points A, B, C, and D, but the points move progressively to the left, because q decreases for each curve as we go up the graph, in the ratio $1:1/2:1/3:1/4$. 'Speeding up' the muscle from point C to point D produces a much smaller gain in power than speeding it up from point A to point B, because point D has moved so far over to the left, that it falls on the steeply rising part of the curve. The fastest muscle becomes inefficient, because the imposed absolute strain rate is so low that it is obliged to operate in what (for it) is almost an isometric condition (at the extreme left side of Fig. 3.3). This muscle is too fast for the task in hand. A slower one produces nearly as much volume-specific power, and does so more efficiently.

This is what is meant by 'matching' the muscle to its load. The mechanical conditions (or 'load') require the muscle to shorten within a limited range of values of the absolute strain rate. For the muscle to perform its function, its own characteristic value of ψ_0, the intrinsic speed, must be adjusted to suit the load, that is, the muscle must not be too 'slow', or too 'fast' either. ψ_0 has to be higher in the swimming muscles of a minnow than in those of a whale, because the smaller animal must beat its tail at a higher frequency, and therefore the strain rate during shortening is higher. If whale muscle were installed in a minnow, it would be too slow to complete a contraction in the short time allowed by the minnow's high tail-beat frequency, and no work would be done. Minnow muscle in a whale would be working under near-isometric conditions throughout the shortening phase. The cycle work would be high, but the efficiency would be very low (Fig. 3.3). Most of the fuel energy consumed by the muscles would be turned into heat, and very little into work.

3.11 Temperature and muscle power

In any particular muscle, ψ_0 increases with temperature, and it is often thought that homoiothermic animals maintain the highest possible body temperature, in order to maximize the power output from their muscles. However large, warm-blooded animals such as vultures must have slower muscles than smaller ones with similar locomotion, like hummingbirds. They cannot 'speed up' their muscles, by temperature or other means, because mechanical constraints do not permit this. ψ_0 has to be adjusted to a value that suits the animal's locomotion, not maximized. An animal whose body temperature is low but constant, such as a fish that lives in constantly cold water, can adjust the value of ψ_0 to suit its tail-beat frequency, at the prevailing temperature. If such a fish is warmed up, its muscles become too fast, and are forced to operate inefficiently at the left-hand end of Fig. 3.3. This is one reason why fish adapted to cold water are notoriously intolerant of temperatures above normal. The need to adjust ψ_0 makes it advantageous to maintain a constant body temperature, but not necessarily a high one. There is a reason why a high body temperature as such is advantageous, but we have to look elsewhere for it.

3.12 Optimizing tonic muscles

Although a tonic muscle is not adapted to do work, ψ_0 is the variable that has to be adjusted, in this case also, to adapt the muscle to its function. The energetic cost for which a muscle maintains force is directly proportional to its particular value of ψ_0 (Section 3.8), but in this case there is no optimum range for ψ_0, as there is for a locomotor muscle. The slower the muscle, the more economically it maintains tension. A ligament is equivalent to an infinitely slow muscle, which can maintain tension for no expenditure of energy at all. The limitation of a ligament is that its length cannot be adjusted. In applications such as the maintenance of mammalian posture, muscles must be used which are fast enough to adjust their lengths as the animal alters its position, but no faster than they have to be. This requirement determines a minimum value for ψ_0 in each particular case. Actually, any muscle can be used to maintain a force, but the cost of using fast muscles in this way is excessive. For example, human fingers are adapted for intricate manipulative tasks, and their flexor muscles therefore have to be fast. They can be used to hang from a window ledge in a moment of emergency, but at the expense of considerable discomfort and energetic cost. Orangutans, however, hang from their fingers for long periods without apparent discomfort, and one must conclude that their finger flexor muscles have lower values of ψ_0 than their human counterparts. Consequently, orangutans are not well suited to tasks that involve rapid finger movements.

3.13 Anaerobic and aerobic muscles

Aerobic muscles are those which oxidize their fuel (usually fat) all the way to carbon dioxide and water, within the fibre. Molecular oxygen has to be supplied to the fibre to achieve this. Anaerobic muscles perform only a partial oxidation within the fibre, and do not require molecular oxygen. The oxidation process in aerobic muscles is carried out in mitochondria. The volume of mitochondria required is related to the rate at which energy has to be processed, hence the proportion of the cell's volume that has to consist of mitochondria is related to the volume-specific power output of the muscle. In muscles with very high specific power output, notably the flight muscles of hummingbirds and various insects, mitochondria can occupy up to half the volume of the cell. This means that only half the cross-sectional area is occupied by contractile proteins, and so the stress is reduced pro rata. The volume-specific power output therefore does not increase linearly with contraction frequency in aerobic muscles, as Equation 3.5 predicts. Pennycuick and Rezende (1984) showed that if the stress–strain product ($\sigma\varepsilon$) of the contractile proteins themselves could be kept constant, the mass-specific power would depend on the 'operating frequency', that is the frequency at which the muscle is adapted to operate, in this way:

$$P_m = \sigma\varepsilon f/\rho(1 + k\sigma\varepsilon f). \qquad (3.10)$$

where k is the inverse power density of the mitochondria, that is the volume of mitochondria required to supply ATP at a rate sufficient to generate 1 W of mechanical power. If the frequency is very low, the expression in the bracket of the denominator is only a little more than 1, and Equation 3.10 is nearly the same as Equation 3.6. If the frequency is very high, the expression in the bracket is nearly the same as $k\sigma\varepsilon f$, and in this case P_m levels off at an asymptotic value of $1/\rho k$. In other words, the specific power output of an aerobic muscle is ultimately limited by the rate at which unit volume of mitochondria can process energy, not by the rate at which the contractile proteins can do work (Fig. 3.7).

This argument does not apply to anaerobic muscles, because they do not require substantial amounts of mitochondria in their fibres, and the stress which they can exert is therefore not reduced for this reason at high contraction frequencies. Anaerobic muscles, at least in vertebrates, are able to develop higher specific power than aerobic ones operating at the same

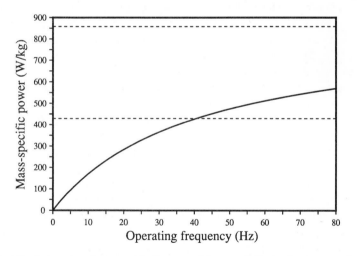

Fig 3.7 The 'operating frequency' of an aerobic muscle is the frequency at which it is *adapted* to operate in cruising locomotion, implying that it contains a volume of mitochondria to suit. Even if the stress–strain product remains constant, the mass-specific power does not increase linearly with operating frequency, because higher-frequency muscles have to contain a greater fraction of mitochondria. The lower dotted line is for muscles in which the volume ratio of mitochondria to myofibrils is 1:1, and the upper one is the theoretical limit for muscles that work at an infinite frequency, with a vanishingly small fraction of myofibrils.

frequency, and the higher the frequency, the greater the advantage. The contractile machinery can also be activated more quickly from a cold start in anaerobic muscles, but, on the other hand, energy conversion is much less efficient. Anaerobic muscles are generally adapted for 'sprint' applications, where short bursts of maximal power are required, whereas aerobic muscles are obligatory for prolonged 'cruising' locomotion, in which oxygen balance has to be maintained, and efficiency is important. Many locomotor muscles contain both aerobic and anaerobic fibres side by side, or contain aerobic fibres which also have some capacity for temporary anaerobic operation. Flying animals commonly use anaerobic metabolism for a short period, when taking off after a period of rest, but change over to aerobic metabolism in prolonged flight.

The rate at which unit volume of mitochondria can process energy is strongly temperature-dependent, so that the higher an animal's body temperature, the less volume of mitochondria is required to supply the energy for contraction, and the higher the muscles' specific power output. This gives an active animal a strong evolutionary incentive to maintain as high a body temperature as possible, not just a constant temperature. Even small insects have to warm up their flight muscles before they can fly, and run them in flight at a temperature comparable with that of mammals and birds. It is difficult for a fish to be warm-blooded, because its gills are just as effective at exchanging heat with the surrounding water as they are at exchanging gases. It would be virtually impossible to keep the whole of a fish's body warm, but some fishes, notably tunas and lamnid sharks, have an elaborate system of retia which permits a part of their body musculature to run at a temperature several degrees higher than that of the water, and of the rest of the body (Carey 1973). The fish selectively warms just the aerobic portion of its body muscles, which is the only part that is active in sustained cruising and is quite a small proportion of the whole. By reducing the volume of mitochondria needed to sustain aerobic metabolism, the higher temperature permits the 'cruise' muscles to occupy less volume than would be necessary in a fully poikilothermic fish, which in return reduces the fish's cross-sectional area, and its hydrodynamic drag. Most of the swimming muscle is white ('sprint') muscle, which is used only for transient bursts of activity. It can be adjusted to run optimally at the water temperature, since it does not require massive amounts of mitochondria. There is therefore no advantage in keeping it warm.

Aerobic muscles vary in colour from pink to deep red, partly because the cytochrome in mitochondria is red when seen in bulk, partly because such muscles usually contain myoglobin to buffer the partial pressure of oxygen within the tissue, and partly because any organ that works continuously has to have a generous blood supply. Anaerobic muscles contain only a few mitochondria and no myoglobin, and can manage with a minimal blood supply, and hence are much paler in colour than aerobic muscles, often white. There is a widespread notion that white muscles are 'fast' and red ones are 'slow', but this an error based on unwarranted generalization from the human condition. Human fast locomotor muscles are primarily adapted to short-period 'sprint' activity, and hence are pale in colour, whereas the slow postural muscles, which operate continuously, necessarily have to be aerobic, and red. The 'speed' of a muscle is expressed by its intrinsic speed (ψ_0), which has no connection with the colour. The fastest vertebrate locomotor muscles, the flight muscles of hummingbirds, are deep red in colour, because they are aerobic.

3.14 Upper limit to strain rate

Fig. 3.7 shows the way in which mass-specific power should increase with contraction frequency, for given values of the stress and strain developed during shortening. Hummingbirds beat their wings at spectacular frequencies (for vertebrates), in the range 15–50 Hz or even more. However, smaller flying animals, such as bees and flies, have wing-beat freqencies in the hundreds of Hz, and tiny midges have been recorded generating sounds up to 2 kHz, presumed to be equal to their wing-beat frequencies. If the volume-specific work (stress times strain) were the same in all these different-sized animals, then the mass-specific power would approach the upper dotted line in Fig. 3.7, but this does not seem to happen. Most estimates of mass-specific power in smaller animals have been obtained by metabolic rather than mechanical methods, and many authors express the power per unit of body mass, rather than per unit of muscle mass. Conversion to mass-specific power in the sense used here is uncertain, as is the conversion of metabolic measurements to the actual rate of doing mechanical work. Ellington (1985) reviewed estimates from metabolic and mechanical lines of evidence, and concluded that the mass-specific mechanical power output of insect flight muscle does not exceed about 200 W kg^{-1}, which is less than one quarter of the asymptotic limit predicted in Fig. 3.7. As flying animals get smaller, the wings have to beat at a higher frequency, for mechanical reasons, but the mass-specific power output of the muscles apparently levels off at a much lower limit than would be predicted by Equation 3.10. The implication is that the cycle work must decrease at high frequencies, rather than remaining constant, as was assumed in deriving Equation 3.10.

Such an effect would result if there were an upper limit to the strain rate. One can imagine that there might well be an upper limit to the frequency with which the myosin cross-bridges can 'walk' along the actin filaments, and this would translate directly into an upper limit on the maximum possible strain rate that any muscle can achieve. The strain rate (ψ) is approximately equal to the active strain (ε) divided by half the period for a complete cycle of contraction and relaxation. If we call the period τ, then

$$\psi = 2\varepsilon/\tau \qquad (3.11)$$

Since τ is simply the reciprocal of the frequency (f), the strain rate can also be written:

$$\psi = 2\varepsilon f. \tag{3.12}$$

In the argument of Section 3.9, which probably is appropriate for large animals including all vertebrates, ε was regarded as fixed, in which case the strain rate increases in proportion to the frequency. However, if the strain rate comes up against a fixed upper limit (ψ_{max}), and cannot increase any further, then the strain has to go down with further increase of the frequency. Turning Equation 3.12 round:

$$\varepsilon = \psi_{max}/2f. \tag{3.13}$$

A s the frequency continues to increase, the strain gets progressively smaller.

Strain is difficult to measure under anything approaching natural conditions, and there are only a few explicit observations in the literature. Stevenson and Josephson (1990) reported that an isolated flight muscle of a large moth *Manduca sexta* produced maximum power when forced to work at a strain of about 8 per cent, at 28 Hz. This works out to a strain rate of $4 \cdot 5 \, \text{s}^{-1}$, from Equation 3.12. Anecdotal evidence suggests that some small hummingbirds increase both wing-beat frequency and amplitude in high-powered manœuvres, as when accelerating away from a flower. There are no observations of muscle strain, but if the strain could reach a maximum of $0 \cdot 25$ as in larger vertebrates, at a frequency of, say, 50 Hz, the strain rate would be $25 \, \text{s}^{-1}$. These assumptions are rather extreme. If there is indeed a fixed upper limit, ψ_{max}, to the strain rate, which applies to all animals, then one could suggest as a 'best guess' that its value is somewhere in the range $10–20 \, \text{s}^{-1}$.

3.15 Difficulties of small animals

If we provisionally take $20 \, \text{s}^{-1}$ as the upper limit for the strain rate, then Equation 3.13 gives $\varepsilon = 10/f$ as a rule of thumb for estimating the strain. A fly beating its wings at 250 Hz would have time to shorten its flight muscles through a strain of only $0 \cdot 4$ per cent at each contraction. Be that as it may, small flying insects certainly cannot get enough excursion in their muscle attachments to crank their wings up and down by direct action, as vertebrates do. They have to resort to indirect mechanical linkages, whereby the muscles distort the thoracic box by a small amount, and this movement is 'geared up' into much larger excursions of the wings. Jumping animals have to perform sudden extreme movements of their limbs, and although this action is not repetitive, it has to be completed in a short time, just as in each contraction of a rapidly beating wing. The smaller the animal, the shorter the time available for extending its legs in a jump. If the extension were done directly by muscle action, the strain rate required in fleas, for example, would be far above any probable limit. Contrary to popular belief, fleas are very poor jumpers, but it is amazing that they can jump at all. I shall return to these interesting pests in Chapter 4.

4. Scaling

4.1 Varying the size but not the shape

We all recognize that our fellow human beings vary quite a bit in size depending, it seems, partly on their individual genetic constitutions, and partly on how well or badly they were fed when young. We also recognize that tiny humans of the 'Tom Thumb' variety are mythical, and so are the fearful giants of old, because they are beyond the extremes of normal variation. Taxonomists pay little attention to size when they classify animals, as they judge it to be useless as a 'character' for inferring ancestry or relationships between species. The implication is that, although a species may have a well-defined, genetically determined size range at any one time, this size range can be moved quite easily, if natural selection favours a race of giants or Tom Thumbs. This seems to be a frequent occurrence, as it is quite usual for a recognizable body form to be shared by a set of related species, which now vary widely in size, although presumably they had a common ancestor in the geologically recent past. For example, dolphins and the great whales are easily recognized as variations on the same body plan, even though they differ in size far beyond the range of variation within any one species.

A major shift in a species' size range would have a number of interesting consequences, as Gulliver discovered. The study of the way in which an animal's characteristics would be expected to change (or not to change) as a result of enlarging or diminishing a particular form to different sizes, is often called 'scaling'. The basic ideas are simple, but sometimes lead surprisingly far when their ramifications are diligently followed. McMahon and Bonner (1983) have written a detailed and highly entertaining account of all aspects of scaling, including the history of the subject back to Aristotle. My purpose in this chapter is to single out certain aspects of scaling which follow from the mechanical thinking of Chapter 3, and which have ecological consequences which will be discussed in Chapters 6 and 7. Some well-known results in scaling need to be rethought in the light of the later discovery of fractal geometry, and I shall look briefly into that in Chapter 5.

4.2 Scaling of simple geometrical shapes

To illustrate the basic principle of scaling, we can compare two squares (Fig. 4.1) one of which has sides of 1 cm and the other 3 cm. The areas are respectively 1 cm^2 and 9 cm^2. If the squares are faces of cubes, then the volumes of the two cubes are respectively 1 and 27 cm^3. The conventions of scaling do not take account of the units in which the measurements are made,

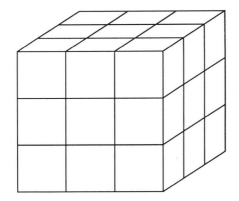

Fig 4.1 Simple geometric scaling — see text.

or even of the dimensions of the original variables (at least not directly). The variables used in scaling are dimensionless *ratios*. We do not consider the actual areas of the two squares of Fig. 4.1, but the *area ratio*, which is 9, and the *volume ratio*, which is 27. The area and volume ratios have these values when the linear ratio is 3. In the short-hand notation of scaling, we say that

$$A \propto \ell^2, \text{ and} \tag{4.1}$$

$$V \propto \ell^3. \tag{4.2}$$

These are spoken as 'area varies with the square of the length', and 'volume varies with the cube of the length'. However, the variables in Proportionalities 4.1 and 4.2 do not have the dimensions of length, area, and volume. The word 'length' is short for 'the ratio of corresponding lengths in two figures of the same shape but different size', and similarly for 'area' and 'volume'. The length of any line, straight or curved, can be used to obtain the length ratio between two figures, and any area or volume will do, regardless of its shape, so long as corresponding areas or volumes are compared. Proportionalities 4.1 and 4.2 work equally well with spheres, if we obtain the length ratio by comparing the equatorial circumferences, the area ratio by comparing the surface areas, and the volume ratio by comparing the volumes. If animals were simple geometrical figures, then we could find the length ratio by measuring the tails of two animals of different size, and then use Proportionality 4.1 to find the ratio of the surface areas of their intestines. Such expectations are sometimes confirmed by measurement, and sometimes not. When they are not, the discrepancy may be due to 'allometry' (below) or to more subtle causes (Chapter 5).

4.3 Testing for geometrical similarity

Underlying most scaling problems is the hypothetical notion of a series of 'geometrically similar' animals, differing from one another in size, but identical in shape. The test of geometrical similarity is that Proportionalities 4.1 and 4.2, or a version of them involving body mass, accurately describe the relationship between length ratios, area ratios, and volume ratios in comparisons between different members of the series. People do not often measure an animal's volume, as would be required to use Proportionality 4.2 for a test, but the mass is easy to measure. It is usually assumed that the density does not vary in different members of a series, and if that is true, then the mass also will vary with the cube of the length, if the animals are indeed geometrically similar:

$$m \propto \ell^3. \qquad (4.3)$$

This relationship is often presented the other way round, that is by saying that the length varies with the one-third power of the mass. For example, if we have a series of birds, and we write m for the body mass and b for the wing span (a length), we expect to find, if they really are geometrically similar, that

$$b \propto m^{1/3}. \qquad (4.4)$$

We can also write S for the wing area. Since any area scales with the square of any length, the wing area should scale with the two-thirds power of the mass:

$$S \propto m^{2/3}. \qquad (4.5)$$

Fig. 4.2 is an example of an 'allometric diagram' designed in this case to test whether the wing spans and body masses of 11 bird species are indeed related by Proportionality 4.4. The species in question are storm-petrels,

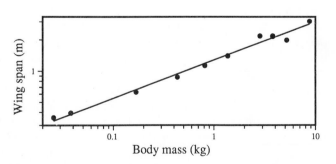

Fig 4.2 Wing spans for 11 species of Procellariiform birds, plotted against body mass. The slope of the standard major-axis line is 0·370 (expected: 0.333).

petrels, and albatrosses belonging to the Order Procellariiformes, a group with a reputation for being biologically 'homogeneous', in spite of covering a 400: 1 range of body mass. The graph is 'double-logarithmic', that is, both the *x* and *y* co-ordinates have been transformed into their logarithms before plotting. The test consists of fitting a straight line through the points and determining its slope. It should be noted that the ordinary linear regression calculation gives a different line, depending on whether *y* is plotted against *x* or *x* against *y*. This is not appropriate in allometric tests, since neither variable is inherently 'dependent' or 'independent'. The slope should be calculated by the 'standard major axis' method of Hofman (1988), also called 'reduced major axis' (Rayner 1985).

To see the meaning of the slope, we can make Proportionality 4.4 into an equation, thus:

$$b = Km^{1/3}, \tag{4.6}$$

where *K* is some constant whose numerical value usually does not concern us. The same equation can be expressed in terms of the logarithms of the variables, thus:

$$\ln b = \ln K + 1/3 \ln m, \tag{4.7}$$

This is the equation of a straight line, whose slope is 1/3. The slope came from the exponent of *m* in Proportionality 4.4. If the measured slope can be statistically distinguished from 1/3, then we have detected a departure from geometrical similarity. Notice that the logarithmic transformation is only permissible because *b* and *m* are dimensionless ratios. The taking of logarithms is an operation that can be performed only on pure numbers, not on quantities with dimensions. It might appear that we have broken this rule in plotting Fig. 4.2, but there is a way around the difficulty. What one does in practice if a bird has a wing span of, say, 0·35 m is to key that number in, and then take its logarithm. This would not be permissible if the '0.35' were a number of metres, with dimensions **L**. You have to think of it as the ratio of the wing span of this particular bird to the wing span of a 'reference bird' which happens to have a wing span of 1 m. That way, the number stays the same, but the dimensions disappear. Strictly speaking, Fig. 4.2 is a plot of the logarithms of span ratios (rather than spans) against the logarithms of mass ratios, although it would usually be described as a 'double-logarithmic plot of span versus mass'.

Tests for geometrical similarity need to be treated with caution. The standard major-axis slope in Fig. 4.2 is 0·370, which is a little higher than the expected value of 0·333, indicating that the larger members of the series have relatively longer wings than the smaller ones. The standard deviation of the slope is 0·0142, from which a *t*-test indicates that the difference from the expected value is significant at the 5 per cent probability level, but only just. Fig. 4.3 shows the wing areas of the same birds, plotted against their

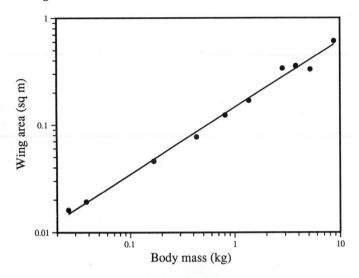

Fig 4.3 Wing areas of the same birds as in Figure 4.2, plotted against mass. The slope of the standard major-axis line is 0·627 (expected: 0·667).

body masses. Here the slope is slightly less than expected, 0·627 as against the expected value of two-thirds. The standard deviation of the slope is 0·0180, and the *t*-test indicates no significant difference between observed and expected values. If we take our statistics too literally, we would conclude

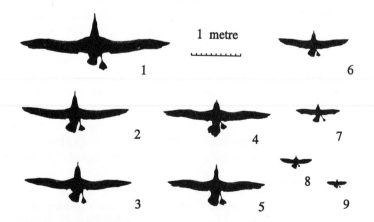

Fig 4.4 Wing tracings of nine of the 11 species in Figure 4.2, 4.3, and 4.6, all drawn to the same scale. The wing span of the largest species, *Diomedea exulans*, is 3·03 m, and that of the smallest, *Oceanites oceanicus* is 0·393 m (from Pennycuick 1982). 1, wandering albatross; 2, black-browed albatross; 3, grey-headed albatross; 4, light-mantled sooty albatross; 5, giant petrel; 6, white-chinned petrel; 7, cape pigeon; 8, dove prion; 9, Wilson's storm petrel.

that these species are very nearly geometrically similar, with a barely significant tendency for the larger ones to have relatively longer wings. Actually,
this would be a highly misleading conclusion.

If we look at the actual wing shapes, it is obvious at a glance that they
deviate in a systematic way from geometrical similarity. Fig. 4.4 shows wing
tracings of nine of the 11 species, all drawn to the same scale, just to show
the wide range of size. The shapes are better appreciated from Fig. 4.5,
in which the scales have been adjusted, so that all of the silhouettes have
the same wing span on the page. The wing shape gets progressively and
unmistakably broader, as one goes from the largest species to the smallest
(top left to bottom right in Fig. 4.5). Scaling wing span and wing area
separately evidently does not tell the whole story. Neither variable by itself
expresses the wing shape, but we can combine them to give a third variable,
the *aspect ratio* (Λ), where:

$$\Lambda = b^2/S. \tag{4.8}$$

This only expresses the shape, being high for a narrow wing, and low for
a broad one, irrespective of whether the wing is large or small. Being the
ratio of two areas, Λ is dimensionless, and expected to be independent of m.

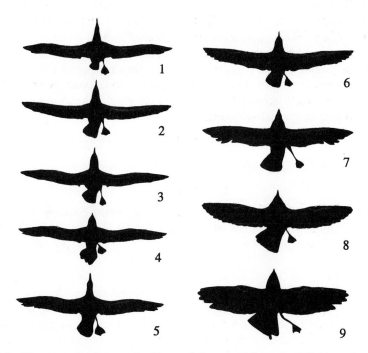

Fig 4.5   The same tracings as in Figure 4.4, enlarged so that all have the same
wing span on the page (from Pennycuick 1982). Species numbers as in Figure 4.4.

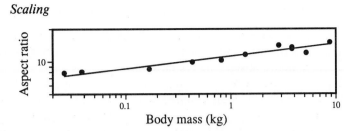

Fig 4.6   Aspect ratio v. body mass for the birds of Figures 4.2–4.5. The slope is 0·116 (expected: zero).

A double-logarithmic plot of aspect ratio v. mass should have a slope of zero. Fig. 4.6 is such a plot and the slope is actually 0·116, with a standard deviation of 0·0111. The data are the same as before, but this time the *t*-test indicates a highly significant difference between the observed and expected results ($P \ll 0·01$). The moral is that if you want your allometric test to be relevant, you have to be careful to test the right variable.

4.4 Scaling work

The notions of scaling were brought to bear on animal locomotion in a classic paper by A. V. Hill (1950). First, Hill noted that the work done by a muscle when it shortens is equal to the force (F) which it exerts, multiplied by the distance (ΔY) through which it shortens. As explained in Chapter 3, the stress which a muscle can exert is essentially a property of the tissue, so that the force is proportional to the cross-sectional area of the muscle. Similarly, for fixed strain, the distance shortened is proportional to the extended length. In scaling terms:

$$F \propto \ell^2, \text{ and } \Delta Y \propto \ell. \tag{4.9}$$

The work done is Q, where:

$$Q = F\Delta Y \propto \ell^3. \tag{4.10}$$

that is, the work done in one contraction by corresponding muscles in geometrically similar animals of different size varies with the cube of the length, or directly with the mass:

$$Q \propto m. \tag{4.11}$$

Even this modest result was enough to allow Hill to consider a dynamical problem. Suppose a series of geometrically similar animals take a standing jump vertically upwards, each using the most work it can get out of its muscles in a single contraction. How will the height of the jump scale? The work done by the muscles is Q, and we suppose that this is all turned into potential energy, raising the animal's weight through a height h:

$$Q = mgh, \text{ or } h = Q/mg. \qquad (4.12)$$

Q, as noted above, scales directly with m, while g, the strength of gravity, is unaffected by m. Therefore h scales with m/m, in other words, it is independent of m. I like to call this 'Hill's First Law of Locomotion': geometrically similar animals all jump the same height (Fig. 4.7). Most people think, on first hearing this, that it must be wildly wrong, but it is not. A fence to contain antelopes needs to be 2 m high, regardless of the size of the antelopes trying to jump over it. The larger species may be able to clear a slightly higher obstacle, but that is because they start with their bellies further off the ground. h is the height through which the animal's centre of gravity is raised in the jump, not the maximum height at the top of the trajectory.

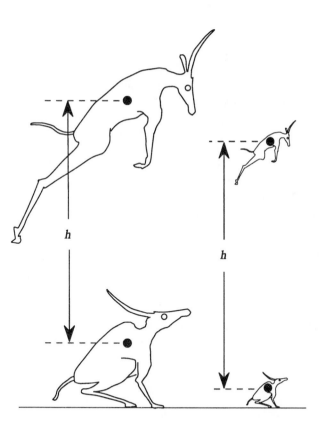

Fig 4.7 Hill's First Law of Locomotion states that geometrically similar animals should all be able to reach the same height in a standing jump. The height h is the distance through which the animal's centre of gravity is raised above its starting position.

4.5 The limitations of fleas

According to the law, if antelopes can jump 2 m, then fleas should be able to do so too. Popular accounts credit the flea with spectacular powers because it can jump many times its own body length, but that is irrelevant. The *absolute* height of the jump should be independent of the mass, not the height relative to the body length. Fleas can jump less than 5 cm straight up, and in the light of Hill's First Law, one has to ask why their performance is so wretchedly inadequate, in comparison with that of mammals. The answer is that even though a flea's muscles may be able to do an amount of work proportional to its body mass, they cannot do it fast enough. It is a simple consequence of Newton's laws of motion that if two animals each take a standing jump to the same peak height as in Fig. 4.7, both of them must leave the ground at the same upward velocity. Each animal accelerates to the same velocity, in the distance allowed by straightening its legs from the fully flexed to the fully extended position. The smaller animal has a shorter distance, and therefore also a shorter time, to accelerate from rest to take-off velocity, and therefore its muscles have to shorten at a higher strain rate.

Bennet-Clark and Lucey (1967), investigating the jumping performance of fleas, estimated that the time available for the legs to straighten in the jump is only about 0.75 milliseconds. Assuming (generously) a mammal-like strain of, say, 25 per cent, that would require a strain rate of $333 \, \text{s}^{-1}$! The estimates of maximum strain rate in Section 3.14 are admittedly rough, but it is very unlikely that any muscle can achieve $30 \, \text{s}^{-1}$, let alone 300. It is amazing that such a small animal can jump at all. As Bennet-Clark and Lucey discovered, it is only able to do so with the aid of a special indirect mechanism. The flea's jumping muscles do not operate the leg directly, but instead distort a piece of rubber-like protein, which acts as a spring, and stores the work done (slowly) by the muscles. When given electric shocks, the flea sets the spring in response to the first stimulus, and is only ready to jump at the second. To jump, it uses a small muscle to operate a trigger mechanism, which releases the spring, allowing the stored energy to be transmitted to the legs. The 'rubber' can change its shape at a much higher strain rate than would be possible for a muscle, quickly enough to hurl the flea into the air. On the other hand, the flea has to carry a heavy piece of elastic protein in addition to its muscles, which is one reason for its modest jumping performance. Air resistance also has a vastly greater effect on a small animal than on a large one, and may even be the major factor in undermining the jumping powers of fleas.

4.6 Scaling frequency

Hill (1950) appears to have been the first to realize that the frequency with which an animal contracts its muscles is generally lower in larger animals than in small ones, and that this in turn determines the mechanical power output that their muscles can deliver, and the rates at which fuel and oxygen are consumed. To find out how the oscillation frequency of a limb scales, Hill considered the angular acceleration produced by a tendon exerting a moment about the basal joint. Assuming that the strength of collagen can be represented by a fixed maximum stress, the force (F) exerted by the tendon should be proportional to its cross-sectional area, and should therefore scale with the square of the length, while the moment arm (y) scales directly with the length. The moment (M) therefore scales with the cube of the length:

$$M = Fy \propto \ell^3. \tag{4.13}$$

In the form of Newton's Second Law that applies to angular motion, force is replaced by moment, mass by moment of inertia (I), and linear acceleration by angular acceleration (j):

$$M = Ij. \tag{4.14}$$

Moment of inertia has dimensions $\mathbf{ML}^2$, and therefore scales with ℓ^5. Equation 4.14, turned around, indicates that the angular acceleration should vary inversely with the square of the length:

$$j = M/I \propto \ell^{-2}. \tag{4.15}$$

Hill then considered the time taken for the leg of a running animal to accelerate through a fixed angle (the step angle), starting at zero angular velocity. By integrating the angular acceleration twice, he showed that this time should be proportional to the length. Thus the period of oscillation of the limb varies directly with the length. The frequency (f), being the reciprocal of the period, varies inversely with the length:

$$f \propto \ell^{-1} \propto m^{-1/3}. \tag{4.16}$$

This result leads to another interesting generalization. The *stride length* (s) of a running animal is defined as the distance between successive footfalls of the same foot. It follows that its speed (V) is equal to the stride-length times the stepping frequency (f). The stride length, like other lengths, varies directly with the length, so:

$$V = sf \propto \ell \times \ell^{-1} = \ell^0. \tag{4.17}$$

That is, top speed is independent of length.

Geometrically similar animals all run at the same top speed. I like to call this 'Hill's Second Law of Locomotion'. Like the First Law, it strikes most people as wrong at first sight. However, Alexander *et al.* (1977) determined the top speeds of 10 species of wild African ungulates, by chasing them with a vehicle. The animals ranged from gazelles to giraffes, with a mass range of 50:1. The allometric plot of top speed v. body mass actually showed that the top speed decreased slightly in the larger animals, although the slope was not significantly different from zero. The larger animals took longer strides, but their lower stepping frequencies compensated for this, as Hill predicted.

4.7 Maximum frequency and cruising frequency

Hill's method of scaling frequency, outlined above, started from the postulate that the tendon causing the limb to accelerate is loaded to some fixed stress, generally assumed to be the maximum stress that it can safely carry, without risk of breaking. The animal is made of materials, including the collagen of tendons, whose strength is limited. Hill's argument identifies a maximum value for the acceleration, and hence for the frequency, beyond which the tendon would be liable to break. If the animal were to try to oscillate its limbs at a higher frequency, and its muscles had the strength to do so, the likely result would be to tear a tendon out of its insertion. This not a purely theoretical hazard — overtrained athletes have been known to do it. The margin of excess strength is no greater than it needs to be for safety in normal circumstances. Hill's scaling argument for speed identifies the animal's *maximum* speed, which is usually not the speed at which it goes about its daily errands.

An animal in prolonged steady locomotion, when migrating, say, or travelling from place to place without undue urgency, will select a comfortable cruising speed that is below its maximum speed. In terms of frequency, there is some *natural frequency*, at which the least amount of effort is needed, just to maintain the oscillation of the limbs. As noted in Chapter 2, for a pendulum of length L, there is a frequency equal to $1/2 \sqrt{(g/L)}$, at which only a very small amount of power has to be supplied in the form of a minute push administered once per cycle, to keep the oscillation going indefinitely. A simple pendulum can also be forced to oscillate at frequencies above or below the natural frequency, but as the frequency deviates from the natural frequency, especially on the high side, the amounts of force and power that have to be supplied to keep it going, increase rapidly.

This notion of a natural frequency also applies in animal locomotion. Alexander and Jayes (1983) showed that in 'gravitational' walking, where the animal behaves like an inverted pendulum, the natural frequency varies inversely with the square root of the leg length, the same law as applies to a normal pendulum. Animals that trot or gallop effectively bounce on springs in the distal parts of their legs, and the natural frequency of this 'pogo-stick' motion also varies inversely with the square root of the length, rather than inversely with the length, as under Hill's 'strength-limited' analysis. The stepping frequencies of wild African animals, observed in the field (Fig. 4.8) varied approximately with the -0.5 power of the shoulder height, as Alexander and Jayes (1983) predicted, and not with the -1 power, as would be expected under Hill's strength-limited argument. The analysis of

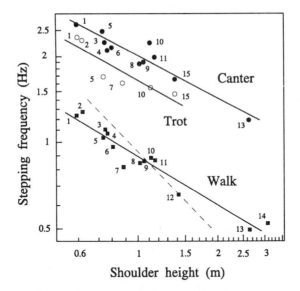

Fig 4.8 The stepping frequencies of wild African animals vary about with the
−1/2 power of the shoulder height, when comparisons are made within any of
the three gaits, walk, trot, and canter. This is the 'pendulum law' predicted by
Alexander and Jayes (1983). A.V. Hill's 'strength limited' law, which requires
the stepping frequency to vary inversely with the length, is not reconcilable with
the data (dashed line). This '−1 power' law is expected to apply to maximum sprint
speed, rather than to cruising speed (from Pennycuick 1975). 1, Thomson's gazelle;
2, warthog; 3, wildebeest (adult); 4, spotted hyaena; 5, Grant's gazelle; 6, impala;
7, lion; 8, hartebeest; 9, topi; 10, zebra; 11, wildebeest (calf); 12, black rhino-
ceros; 13, giraffe; 14, elephant; 15, buffalo.

the wing-beat frequency of birds in Chapter 2 leads to the same conclusion.
It implies that, for geometrically similar birds, the wing-beat frequency
would vary inversely with the square root of the length. As in all cases where
geometrical similarity is assumed, this is the same as saying that frequency
varies inversely with the one-sixth power of the body mass:

$$f \propto \ell^{-1/2} \propto m^{-1/6}. \tag{4.18}$$

As a working hypothesis, I shall assume that this rule applies to any kind of
walking or flying animal in *cruising* locomotion. Proportionalities 4.16 and
4.18 do not tell us the absolute values of the maximum or cruising frequencies for any particular animal, but they do tell us what the *slopes* of the lines
would be, if these two frequencies were plotted against body mass in an
allometric graph like Fig. 4.9. For an animal to be capable of cruising
locomotion, its maximum frequency obviously must be higher than its
natural frequency. Because the slopes of the lines are different, they

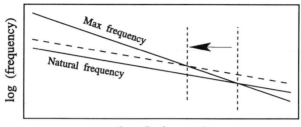

log (body mass)

Fig 4.9   Maximum stepping or flapping frequency, determined by Hill's argument, is not affected by an increase in the strength of gravity, but the natural frequency for cruising locomotion increases in proportion to $\sqrt{g}$ (dotted line). The maximum mass of an animal that is strong enough to run or fly at its natural frequency therefore decreases if gravity increases.

converge towards the right of the diagram, that is in the direction of increasing mass. This implies that for some value of the body mass, the natural frequency is equal to the maximum frequency, and that animals with still higher body masses would not be able to run at their natural frequencies without exerting forces sufficient to break their muscle insertions. This argument tells us that there is an upper limit to the mass of *geometrically similar* animals that are capable of cruising locomotion, although it does not supply a numerical value for the maximum mass.

For an animal that is subject to selection pressure to become still larger, the easiest way to avoid the difficulty shown in Fig. 4.9 is to deviate from geometrical similarity, by making the muscle insertions relatively thicker than in smaller animals of similar type. Real series of animals, such as cursorial mammals or flying birds, invariably do depart from geometrical similarity in such ways. However, it was noted above that a very small departure from geometrical similarity, so small as to be difficult to demonstrate statistically, leads to gross and obvious changes of shape when extended over a wide range of body mass. It is often possible to 'bend' a limit like that of Fig. 4.9, but only at the expense of distorting the animal to a degree that soon makes it mechanically impractical, for reasons not directly connected with the original problem. Fig. 4.9 represents a 'limit law' and it is a real limit, although the cut-off may not be as clean as the diagram would suggest.

It was noted in Chapter 2 that the tail-beat frequency in fishes of different size is expected to vary with ℓ^{-1}, rather than with $\ell^{-1/2}$, as indicated by Proportionality 4.18 for walking and flying animals. This is the same as Hill's strength-limited law for maximum speed. Therefore the cruising frequency would not be expected to converge on the maximum frequency as the size of swimming animals increases, and this particular consideration would not lead to an upper limit to their size. Actually there is no known mechanical limit to the size of swimming animals, and it is not even clear whether there is any limit. If there is one, it may be due to physiological rather than mechanical factors. For example, the time taken for blood or nerve signals to travel from one end of the body to the other could cause problems for a very large animal (Section 4.9), or excessively large size could lead to an unacceptably low reproductive rate (Section 4.13).

4.8 Varying gravity

According to the known laws for walking and flying animals, the *natural* frequency varies with the square root of gravity, in addition to its dependence on body length (or mass). The *maximum* frequency according to Hill's strength-limited argument is not affected by gravity. An increase in the strength of gravity will have the effect of raising the line for natural frequency, while leaving the one for maximum frequency unchanged (Fig. 4.9). The intersection of the two lines, defining the maximum mass, will move to the left (lower mass). A small increase in gravity has a large effect on the limiting mass for a walking or flying animal, at which it is possible to run or fly without overstressing the structure. However, this does not apply to swimming animals, whose weight is supported hydrostatically by the water.

Future naturalists describing the faunas of big planets, whose gravity is stronger than ours on Earth, may anticipate that the largest walking and flying animals will prove to be smaller than those at home, although this may not be true of the swimming ones. Conversely, planets with weak gravity should teem with running and flying animals of great size. It is an interesting coincidence that at two particular periods in the past, in Cretaceous and Miocene times, walking animals occurred that were considerably larger than any living or recent forms, and at the same two periods, the largest flying animals (a pterosaur in the Cretaceous and a bird in the Miocene) were much larger than any living forms. The Miocene bird (*Argentavis*) was morphologically very like a condor (Campbell and Tonni 1983). Its humerus was about twice as long as a condor's, and therefore, if it scaled isometrically, *Argentavis* would have had eight times a condor's mass. As modern condors seem to be only marginally capable of flapping flight, *Argentavis* would be easier to understand if gravity had half its modern value in Miocene times. The earth expansion hypothesis of Carey (1976) serves as a reminder that it is unsafe to assume that surface gravity (or any physical characteristic of the earth) has remained constant over geological time. *Reduced* gravity in the past is difficult to reconcile with Carey's scenario, but on the other hand, the existence of giant animals at various times in the past is a fact that has to be accounted for.

4.9 Scaling power

The power output (rate of doing work) described in Fig. 3.2 was found by multiplying the force exerted by the speed of shortening. In steady locomotion, the *average* rate of doing work is only about half of this amount (or less), because the muscle works intermittently, and spends around half of its time being passively extended, in readiness for the next contraction. Most of the more interesting ramifications of Hill's (1950) analysis follow from his recognition that the average power output of a cyclically contracting muscle can be found by multiplying the cycle work by the contraction frequency. Scaling arguments can be applied to both of these variables. The cycle work varies directly with the mass (Proportionality 4.11), and Hill's strength-limitation argument called for the contraction frequency to vary inversely with the one-third power of the mass:

$$Q \propto m, \text{ and } f \propto m^{-1/3}, \tag{4.19}$$

whence the power output varies with the two-thirds power of the mass:

$$P = Qf \propto m^{2/3}. \tag{4.20}$$

Hill used this result to make deductions about rates of fuel and oxygen consumption, and heat production, although these apply to cruising rather than to top-speed locomotion. I shall summarize some of his results, but some of them have to be modified, as it now seems that the contraction frequency in cruising locomotion usually varies inversely with the one-sixth (rather than the one-third) power of the mass (Section 4.7). Proportionalities 4.19 are replaced by:

$$Q \propto m, \text{ and } f \propto m^{-1/6}, \tag{4.21}$$

so that:

$$P = Qf \propto m^{5/6}. \tag{4.22}$$

For the moment, we stay with Proportionality 4.20, so as to follow Hill's original train of thought. One might imagine, as many people do, that the power output of an animal's muscles would be directly proportional to their mass, but according to Proportionality 4.20 this is not so. The power output is proportional to the animal's *surface area*, rather than to its volume. The power output per unit mass of animal is less in large animals than in small ones.

Hill's next step was to say that fuel and oxygen for locomotion are consumed, and heat produced, at rates that are proportional to the mechanical power output of the muscles. The rates at which different animals require oxygen therefore vary with the two-thirds power of the mass. So do the surface areas of their lungs, so the rate at which oxygen has to be absorbed through each unit of surface area remains constant. Similarly the surface area available for heat dissipation scales in the same manner as the rate of production of heat, so that the heat flux (rate of heat flow through unit surface area) does not change in different-sized animals. The rate at which food is required is likewise proportional to the cross-sectional area of the hole through which it has to enter, and to the surface area of the teeth that grind it up, and of the gut that absorbs it. The stroke volume of the heart is directly proportional to the mass, but the rate at which it is required to pump blood varies only with the two-thirds power of the mass, to match the rate at which the locomotor muscles require fuel and oxygen. Therefore the heart's contraction frequency should vary inversely with the one-third power of the mass, as in the locomotor muscles. The rate at which the heart supplies blood (volume/time) varies with the two-thirds power of the mass, and it pumps the blood into a pipe (the aorta) whose cross-sectional area also varies with the two-thirds power of the mass. The velocity with which the blood travels along the aorta therefore does not vary in animals of different size. A blood corpuscle travels along a mouse's aorta at the same absolute velocity as along an elephant's. The time taken for a blood corpuscle to make a circuit of the animal and return to the heart is longer in the elephant, as it varies directly with the length, or the one-third power of the mass. There is more, but I will not spoil the reader's enjoyment by recounting further ramifications. Hill's great paper deserves to be read in the original.

4.10 Empirical approach to scaling energy consumption

The empirical way to measure the rate at which an animal consumes energy when running at a steady speed is to make it run on a treadmill and measure the rate at which it consumes oxygen. Heglund and Taylor (1988) reviewed a large volume of treadmill experiments on animals ranging from mice (30 g) to horses (200 kg), and found that the 'preferred' speed for each animal varied with approximately the $0 \cdot 2$ power of the mass (predicted $0 \cdot 17$), while the stepping frequency at that speed varied with the $-0 \cdot 15$ power of the mass (predicted $-0 \cdot 17$). Analysing the data on oxygen consumption, they concluded that the energy consumed in each contraction is directly proportional to the mass of muscle. Thus they arrived at Proportionality 4.11, 38 years after Hill (1950) used it as the *starting point* of his argument. Suffice it to say that, if Hill's argument is modified by assuming that the stepping frequency in cruising locomotion varies with the $-1/6$ (instead of the $-1/3$) power of the mass, as predicted by Alexander and Jayes (1983), then the empirical results agree very well with the theory.

There is an extensive literature on 'basal metabolism', meaning measurements of the oxygen consumption or heat production of animals sitting quietly in cages (McNab 1983). It has long been known that allometric plots of the rate of energy consumption v. body mass show slopes less than 1, when comparing related animals of different size. Usually the slope is about $0 \cdot 75$.

This is exactly mid-way between the two-thirds slope deduced by Hill, and the five-sixths slope according to Proportionality 4.22. However, 'basal metabolism' is measured in inactive animals, and has no obvious connection with the power requirements for locomotion. As these measurements were first made on mammals, it was thought that the rate of production of heat was determined by the animal's need to maintain a constant body temperature. If the rate of heat production is to be equal to the rate of heat loss, which in turn is proportional to the surface area of the body, this would give a rough explanation for the allometric slope. However, it later transpired that poikilothermic fishes, amphibians, and reptiles showed approximately the same slope when the comparison was made within one of these groups. Poikilothermic reptiles, for example, produce a line that is parallel to the line for mammals (same slope), but lower, meaning that a poikilothermic reptile consumes energy at a lower rate than a homoiothermic mammal of equal mass. McNab (1983) goes into the conditions that determine whether or not an animal can be homoiothermic, and the reasons why some animals are homoiothermic in some circumstances and poikilothermic in others.

The '0·75 power law' turns up very persistently in allometric studies of rates of energy consumption, seemingly regardless of the group of animals concerned, or the exact purpose for which the energy is being processed. I shall extrapolate a little on this basis in Section 4.13, and make use of the result in Chapter 6.

4.11 Scaling 'Cost of Transport' and Performance Number

Both physiologists and engineers are interested in comparing the 'economy' of getting from place to place by different methods of locomotion. Schmidt-Nielsen (1972) introduced a variable which he called the 'cost of transport' for making such comparisons between different animals, and between animals and machines. Cost of transport was defined as the energy required to move unit weight of animal unit distance. The energy measures available to Schmidt-Nielsen were indirect estimates of the *rate* of energy consumption, derived from measurements of the rate of oxygen consumption of an animal running, swimming, or flying at some steady speed (V). We can call this rate of energy consumption (or power) P_m, where the subscript m stands for 'metabolic' power — that is the total rate at which the animal consumes chemical energy for all purposes, as opposed to an unsubscripted P, which represents the mechanical power produced by the muscles. Schmidt-Nielsen's 'cost of transport' (C_m) was defined as

$$C_m = P_m/WV, \tag{4.23}$$

where W was defined as the animal's weight. The units were cal g^{-1} km^{-1}, indicating that although 'weight' was mentioned in the definition, the results were actually being calculated from body mass, not weight. It is a better idea to use the weight, and write Equation 4.23 more explicitly as:

$$C_m = P_m/mgV, \qquad (4.24)$$

where m is the body mass, and g is the acceleration due to gravity. The numerator of the right-hand side of Equation 4.24 is power (energy/time), while the denominator is force multiplied by velocity, which has the same dimensions ($\mathbf{ML^2 T^{-3}}$). Thus C_m is a dimensionless number, and needs no units. If mass were used in place of weight in the denominator of Equation 4.24, C_m would have the dimensions of acceleration, which is by no means so useful. It is better still to use P rather than P_m in the numerator. P_m is a physiological estimate of the rate at which fuel energy is consumed by the animal, not only to power the muscles, but also for basal metabolism and any other energy-consuming process that may be going on, whereas P is the actual rate at which the muscles are doing *mechanical* work. In some experiments, P can be measured directly from the force and movement at muscle insertions, but more often it has to be estimated, by first subtracting the basal metabolic rate from P_m, and then multiplying by the conversion efficiency, usually assumed to be around $0 \cdot 23$. This gives an unsubscripted cost of transport (C) whose value is between a quarter and a fifth of that of C_m, thus:

$$C = P/mgV. \qquad (4.25)$$

C is the ratio of two powers which are both purely mechanical in nature, without any complications due to the conversion of fuel energy into work. Mechanical work has to be done at a rate P to push the animal forward at the speed V, which is equivalent to saying that the animal is exerting an average thrust force T, directed horizontally forwards, whose magnitude is $T = P/V$. Equation 4.25 then becomes:

$$C = TV/mgV = T/mg. \qquad (4.26)$$

'Cost of transport', if defined carefully in this way, is the ratio of the average horizontal force, needed to push the animal along, to its weight. It explicitly involves gravity.

The reciprocal of this 'cleaned-up' version of the cost of transport is a very well-known variable in aeronautics, the ratio of lift to drag, whose significance was well understood by the early aviation pioneers. The Wright brothers' success followed a systematic experimental search for wing shapes with a high ratio of lift: drag. For flying machines and animals one can write:

$$N = L/D = 1/C. \qquad (4.27)$$

N is not exactly a lift: drag ratio in walking and swimming, but if we call it a 'performance number', and define it as the reciprocal of C in

Equation 4.25, it will serve to compare any form of locomotion with any other:

$$N = mgV/P. \tag{4.28}$$

The ratio of speed to power is not fixed for a given type of locomotion, but is liable to vary at different speeds. There may or may not be some identifiable speed at which the performance number is at a maximum. In flying animals there is a fairly well-defined speed known as the maximum range speed (V_{mr}), at which N passes through a maximum. An animal in prolonged cruising flight, such as a migrating bird, has little option but to choose a speed near V_{mr}, if it is to arrive at its destination. Theoretically, it is also possible to find a corresponding speed for swimming animals, but the speed for maximum N is too slow to be used for cruising. In swimming animals, as in ships, it is always possible in practice to go further by slowing down. Walking and running are the most difficult forms of locomotion to deal with theoretically, but there is a large body of experimental information, from oxygen consumption measurements, which indicates that N varies very little over a wide range of speeds that can be used for cruising, and only declines noticeably at high sprint speeds. Much of this material has been summarized by Heglund and Taylor (1988).

When considering how the performance number should scale in animals of different size, flying animals can be compared on the basis that each animal flies at its own V_{mr}, thus maximizing its performance number. Alexander and Jayes (1983) have shown that geometrically similar walking or running animals can be compared when each proceeds at a speed proportional to the square root of its leg length, so as to give a constant value to the Froude number (F), defined as

$$F = V^2/g\ell. \tag{4.29}$$

The Froude number is an example of a 'similarity criterion'. Its significance is that if two animals of different size run at speeds such that the Froude number is the same for both, then their motions are 'similar' in some identifiable sense. For example, mammals of different size change their gait from a walk to a trot at a fixed value of the Froude number. It is not so easy to identify a similarity criterion for comparing locomotion in different swimming animals, but a reasonable method of doing it can be found in Pennycuick (1987). The conclusion from that particular analysis is that for geometrically similar animals that either swim or fly, the performance number should be independent of body mass, whereas for those that walk or run, it should vary with the one-third power of the mass. Fig. 4.10 is redrawn from Schmidt-Nielsen (1972), with the original y-scale on the left-hand side, and an absolute 'performance number' scale on the right, calculated on the (insecure) assumption that the conversion efficiency was 0·23. The cost of transport for all three forms of locomotion (running,

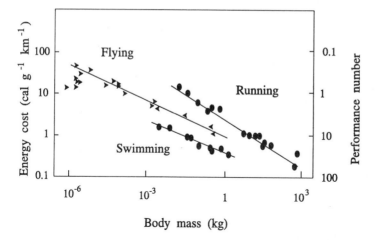

Fig 4.10 'Cost of Transport' in running, swimming, and flying animals, after Schmidt-Nielsen (1972), whose original scale of units is on the left. The dimensionless 'performance number' scale on the right is based on the assumption that the conversion efficiency is 0·23 for all the observations.

swimming, and flying) declines quite steeply with increasing body mass. As noted above, this decline of cost (or increase of performance number) is expected for running, but not for swimming or flying. The slopes are less steep for swimming and flying than for running, but still steep enough to be difficult to explain.

Thinking in terms of performance number raises suspicions that these much-quoted data may not have been strictly comparable with each other. If one looks at the right-hand scale, the larger flying animals (birds) show performance numbers (that is, effective lift: drag ratios) around 10–20, which is in good general agreement with estimates based on mechanical considerations. In this size range there are also secondary allometric and aerodynamical effects which account for the slope of the line (Pennycuick 1989). The cluster of points at the left, however, representing insects, shows performance numbers between 0·1 and 1, which is not so easy to reconcile with any conceivable mechanical scenario. An animal which could do no better than that would have a strong selective incentive to try some other method of locomotion. There are several possible ways in which the 'cost' points for the smaller animals could have been biased upwards. The ratio of power to speed was not observed as a function of speed, but was calculated by comparing separate observations of power and speed. Some at least of the observations of power in insects were made on animals hovering in a confined space, where the power is likely to be higher than that in cruising flight. Also some physiologists believe that the conversion efficiency is lower in insects than in flying vertebrates, and declines progressively with decreasing size. Both of these effects would 'tilt' the line in the observed direction. The line for swimming animals could also have acquired its slope if the animals were not observed in comparable 'cruising' locomotion. The criteria for making this comparison in swimming animals are arguable, and as Schmidt-Nielsen's data were culled from published observations by many different authors, no consistent criterion could be applied, or even discussed.

4.12 Performance Number and gravity

The dependence of performance number on gravity is not strictly a matter of scaling, but it has an interesting bearing on the interpretation of Fig. 4.10. As noted in Chapter 2, it would be difficult for a fully submerged animal to detect a change in gravity. The power required by an animal to swim, in a medium whose density is nearly the same as its own, is determined by the stress in its muscles, and the speed, but is not affected by gravity. Therefore, whatever criterion is used to define the speed for 'cruising' locomotion, the performance number (mgV/P) is directly proportional to gravity, because m, V, and P are all independent of gravity. In both walking at a fixed Froude number, and flying at the maximum range speed, the speed varies with $g^{1/2}$, whereas the power varies with $g^{3/2}$. From Equation 4.28, this makes the performance number independent of gravity. The effect on Fig. 4.10 is that if gravity is increased, the line for swimming would move downwards (higher performance number), whereas the lines for running and flying would stay put. The fact that the 'swimming' line is quite near the other two lines should be seen as yet another of these endless coincidences that make our home planet a favourable place for life. On a planet with gravity much lower than Earth's, the 'cost of transport' for swimming would be higher than for walking or flying, whereas on a big, high-gravity planet, the reverse would be the case. Creatures evolving on a low-gravity planet would more readily take to the air, whereas those evolving in high gravity would be quick to take to the water, or reluctant to leave it in the first place.

4.13 Scaling reproductive rate

Ecologists often discuss the energy relations of organisms in terms of an 'energy budget'. For example, a herbivorous mammal such as an antelope eats vegetation for several hours each day, and extracts Gibbs free energy from it at some average rate which we can call P_i. This is its *power income*. Just as a person's money income is expressed as the amount received per hour, per week or per year, so an animal's income is the rate at which it receives energy. P_i has the dimensions of power ($\mathbf{ML}^2\mathbf{T}^{-3}$). The animal also has a *power expenditure* (P_e), which is the average rate at which it expends energy for such unavoidable purposes as basal metabolism, chewing and digesting food, keeping warm, defending its territory (if it has one), and keeping clear of predators. The difference between the animal's income and expenditure is its *power surplus* (P_s):

$$P_s = P_i - P_e. \tag{4.30}$$

If the feeding is good, and other pressures moderate, the animal will receive energy faster than it expends it, making P_s positive. Gibbs free energy comes in the form of energy-rich compounds obtained from food. To have an income of Gibbs free energy, the animal has to take in material, while an expenditure means that material (fuel) is consumed and the waste products eliminated. Thus a power surplus, whether positive or negative, is mirrored by a corresponding material surplus. If an animal gains (or loses) energy, it also gains (or loses) mass. An animal that maintains a positive power surplus over a period of days or weeks either grows or gets fatter. One with zero power surplus just maintains a constant body mass, while if the power surplus is negative the animal has to live on its reserves, and eventually becomes emaciated, starves, and dies. Ecologists often treat changes of mass as being equivalent to energy gain or loss, because changes of animal or plant biomass correspond in a general way with changes of stored energy, and are easier to measure.

A positive power surplus, maintained over a substantial period, is the essential prerequisite for a female mammal to maintain a growing foetus, and then supply milk to the infant after it is born. Over one or more generation cycles, reproduction in any organism requires a positive power surplus, on average. One can apply the notions of power income, expenditure, and surplus to populations of organisms, as well as to individuals. A positive power surplus in a population of a species can have two components, which may occur separately or together. The surplus can reflect an increase in the number of individuals due to reproduction, or an increase in the average mass of an individual, due to growth or the accumulation of reserves, or some combination of the two.

Various components of power expenditure (P_e) were considered above in connection with Hill's (1950) paper on scaling, with the conclusion that the rate of energy expenditure, in similar animals of different size, is never proportional to the body mass of the animal, but varies with some power of the mass between 2/3 and 5/6, typically $0 \cdot 75$. Hill came to a similar conclusion about P_i, the power income, which he deduced should vary with $m^{2/3}$, although in practice this relationship too is probably nearer to $m^{0 \cdot 75}$. It is a reasonable surmise that *all* powers associated with physiology, metabolism, and locomotion vary roughly with the $0 \cdot 75$ power of the mass. At any rate no exceptions are known. If the power surplus that an animal can make in good feeding conditions varies with $m^{0 \cdot 75}$, then so would the rate at which energy is made available for reproduction. Suppose that we can write for the power surplus of an individual animal:

$$P_s \propto m^{0.75}. \tag{4.31}$$

Then if we imagine a population of N animals peacefully grazing, and all making the same power surplus, we can write that the power surplus for the population ($P_{s.pop}$) is N times that for an individual:

$$P_{s.pop} = NP_s \propto Nm^{0.75}. \tag{4.32}$$

This population power surplus is the power supply for reproduction. If we neglect the fact that young animals initially have less mass than their parents, and average the rate of increase of numbers (dN/dt) over several generations, then we can postulate that this rate of increase is proportional to the population power surplus, that is:

$$dN/dt \propto P_{s.pop}/m \propto Nm^{-0.25}. \tag{4.33}$$

The reader will recognize the familiar 'exponential growth' relationship for a population unrestrained by density-dependent effects on mortality or fecundity. The rate of increase of population number is proportional to the number already in the population, but the constant of proportionality is itself a function of the body mass. Large animals have lower reproductive rates than similar but smaller ones. Western (1980) made an allometric plot of the reported reproductive rates of African mammals ranging from an elephant shrew (70 g), to the African elephant (2580 kg), and found a slope of -0.327, slightly steeper than suggested in Proportionality 4.33. Western's estimate agrees more closely with Hill's (1950) prediction that power should vary with the 2/3 power of the mass, than with the more frequently observed '0.75 power' law.

As we transfer attention to systems of increasing scale, first from the cellular level to whole animals, then to ecosystems, it becomes progressively more difficult to describe processes in terms of simple physical variables. The 0.75 power law provides a bridge between the whole-animal level and the ecosystem level, as I shall attempt to show in Chapter 6, and leads to some interesting insights into the way in which ecosystems regulate themselves, and respond to disturbances. First, in Chapter 5, I have to make a diversion to show how some of the classical ideas of scaling have had to be revised, as a result of the recent discovery of fractal geometry.

5. Fractal objects

5.1 'Length' of an irregular line

Common sense indicates that anything elongated should have a 'length', measurable in metres or some units with dimensions **L**. An animal's intestine, for example, may form a compact lump when in its normal position, but theoretically its length could be measured if it were unravelled. However, some 'lengths' resist measurement, even in theory. This was brought to light by Richardson (1961), who was interested in the origins of wars. Historically, European wars have often been related to border disputes, and so it happened that Richardson was studying the border between Spain and Portugal. He referred to a Spanish encyclopaedia, which said the border was 987 km long, and to a Portuguese encyclopaedia, which said it was 1214 km long. Most of us would have put this down to inaccuracy or error, and thought no more about it, but Richardson tracked the discrepancy to its source, and discovered something much more interesting. Ancient land borders are usually irregular lines, because they follow natural features like ridge lines and rivers. To measure the 'length' of such an irregular line on a map, what you do in practice is set a pair of dividers to some suitable small interval, and step with it along the border. If the step length is s, and n steps are required to traverse the line from end to end, then you estimate the length of the line as ns. This procedure gives a unique value for the length of a straight line, independent of the step length. If you take shorter steps, you have to take more of them, and ns is the same as before. However, if you step along an irregular line with shorter steps, you follow around some small kinks which were previously bridged by the longer steps, and so your estimate of the total length increases. Richardson discovered that the Portuguese cartographers had stepped along the common border using shorter steps than their Spanish colleagues, and so had arrived at a greater total length for the border.

One might suppose that by progressively shortening the steps, the estimated length of the line would eventually level off at the 'real' value. So it does with a smooth curve such as a circle. If you measure the circumference of a circle with a step length equal to the radius (r), say, the estimate is far too low, but if the step length is progressively reduced, the estimated circumference quickly approaches a limit at $2\pi r$. When Richardson tried this with borders and coastlines, using progressively smaller steps, on maps of progressively larger scale, he found that the estimated length went on increasing, without showing any sign of levelling off. He concluded that irregular lines of this kind do not have an identifiable 'length'. You

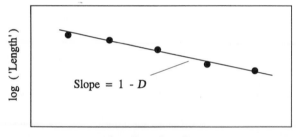

Fig 5.1 The 'length' of an irregular line such as a coastline, as estimated by stepping along it on a map, depends on the length of the steps used to measure it.

can make the 'length' as large as you want, by suitable choice of the step length s. Richardson expressed this by measuring the 'length' of the same stretch of border or coastline, using steps of different lengths, and plotting the estimated total 'length' against the step length on a double-logarithmic graph in the manner of Fig. 5.1. The result is generally a straight line, sloping downwards. For reasons that will be clear in a moment, I shall call the slope 1–D. For a smooth curve like a circle, the line eventually becomes horizontal when the step length is small, that is, the slope is zero, and $D = 1$. For coastlines, the slope is negative ($D > 1$), and the more rugged the coastline, the steeper the slope. Typical values of D for coastlines range from 1·1 for a smooth one to 1·5 for a rugged one.

The problem was taken up by Mandelbrot (1967), who addressed the following question: if a coastline does not have a length, then what property does it have, that expresses how much of it there is, can be expressed as a single number, and does not change when you vary the step length? Mandelbrot's solution was that it has an 'extent' (E), defined as

$$E = ns^D, \tag{5.1}$$

where n is the number of steps as before, and D can be determined as above, from the slope of a graph like Fig. 5.1. E, so defined, does not vary as the step length is changed. Since n is a dimensionless number, and s has dimensions $\mathbf{L}$, the dimensions of 'extent' are $\mathbf{L}^D$. D is called the 'fractal dimension'. It is not usually an integer.

5.2 Mandelbrot's fractal geometry

From this beginning, Benoit Mandelbrot performed the amazing feat of creating a whole new branch of mathematics, which he called 'fractal geometry'. It deals with curves and surfaces that are inaccessible to the methods of calculus, because they cannot be differentiated. The formal science of fractal geometry is primarily concerned with artificially constructed lines and surfaces, but Mandelbrot has always emphasized that the original inspiration came from nature. In his famous book *The Fractal Geometry of Nature* (1983) he explains the basic principles in a way that anyone can understand, and goes on to show how these principles can be used to describe and analyse a tremendous variety of fractal objects, many of which have clear links with physical and biological problems. The startling illustrations in the book also started a new artform of computer-generated fractal images. Basically, the images are abstract, but many have an uncanny resemblance to natural objects. Nowadays, no microcomputer's software library is complete without a 'fractal landscape generator'.

Figure 5.2 is a simple example of an artificially constructed 'coastline', the Koch island, or snowflake. The construction begins with an equilateral triangle (Fig 5.2A). Then, the middle one-third of each side is cut out, and replaced by two sides of a smaller equilateral triangle, whose sides are one-third as long as the original sides, giving the six-pointed (12-sided) star of Fig. 5.2B. The middle one-third of each of the 12 sides is then cut out and replaced by two sides of a still smaller equilateral triangle (Fig. 5.2C), and so on indefinitely. The curve is said to be 'self-similar', meaning that if you take a small segment of the circumference and magnify

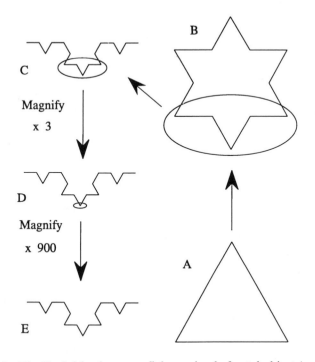

Fig 5.2   The Koch island or snowflake, a simple fractal object (see text).

it, it looks the same, irrespective of the scale of the magnification (Fig. 5.2D, E). It is this property of self-similarity that distinguishes fractal objects from those which are not fractals.

In real physical or biological objects, self-similarity can only be maintained over a finite range of scales, but not so with mathematical objects like the Koch island. The 'circumference' of this object, as measured by stepping around it, obviously depends on the step length. Each time the step length is reduced by a factor of three, you have to step around four short sides, which previously were stepped over with one step, of only three (not four) times the length. The 'circumference' therefore increases by a factor of 4/3. As the step length is made ever shorter, the 'circumference' goes on increasing by a factor of 4/3, each time the step length is shortened by a factor of three. It never levels off. Mandelbrot shows that the fractal dimension (D above) is log(4)/log(3), or about 1.26. Although no meaning can be attached to the 'circumference' of a Koch island, it is possible to distinguish a big Koch Island from a small one, by finding the fractal extent (E) of the boundary. If n steps of length s are needed to step around the island, then:

$$E = ns^{1.26}. \tag{5.2}$$

For any particular Koch island, E does not change if the step length is changed (unlike the 'circumference'). The extent of the boundary (E) is a unique number, and therefore can be used as a measure of the size of the Koch island.

5.3 Measurement of fractal extent

For the ecologist who may actually wish to measure the extent of a coastline, there remains a practical problem which is not addressed by the mathematicians. What is the appropriate SI unit to measure something whose dimensions are $L^{1.26}$? Evidently the answer is the metre$^{1.26}$. This makes it difficult to compare the nest densities of, say, bald eagles, along the coastlines of two islands with different fractal dimensions, a problem which was addressed by Pennycuick and Kline (1986). First of all, to streamline the nomenclature, we proposed a unit called the 'metron', which is really a class of units rather than a single unit. A metron must have a dimension associated with it. A metron of dimension 1 is the same as a metre, and is appropriate for measuring the lengths of smooth lines, while a metron of dimension 2 is the same as a square metre, and can be used for measuring the areas of smooth surfaces. The metre and square metre are no longer seen as distinct and unconnected units, but rather as members of a continuous spectrum of metrons. Metrons of dimension between 1 and 2 can be used to measure the extent of irregular lines such as coastlines. An irregular line can be thought of as having a physical character intermediate between 'length' and 'area'. A line so infinitely convoluted that it leaves no empty space at all would have dimension 2, and would be the same as an area. Typical irregular lines are part-way to that extreme, and have dimensions between 1 and 2.

5.4 Fractal surfaces

Just as one can attempt (unsuccessfully) to measure the length of a coastline by stepping along it, so one could try to measure the area of a mountainside by 'tiling' it with flat tiles, all having the same area, and adding up the total area of the tiles. If the operation were now repeated with smaller tiles, it would be possible to follow around humps and gullies that were bridged over with the larger tiles, and so the total area would increase. The 'area' of a rugged surface is an undefined quantity, just like the 'length' of an irregular line. It goes on increasing without limit, as the area of the measuring tiles is reduced. A rugged surface is physically intermediate in character between a smooth surface and a volume, and can be represented as a fractal of dimension between 2 and 3. Metrons with dimension between 2 and 3 are appropriate for measuring the extent of rugged surfaces. A metron of dimension 3 is the same as the cubic metre, the unit of volume.

Figure 5.3 represents a simple way of measuring the extent and dimension of an island's *surface*, as opposed to its coastline. This can be done from a contour map, which represents the island as a pile of slabs, each h metres thick, where h is the vertical spacing between the contour lines. The dimension of the *boundary* of each slab can be found in the same way as for the coastline, by stepping around the contour with steps of different length, and finding the dimension D from the slope of a line like that in Fig. 5.1. The calculation is easiest if D does not vary too much from one level to the next, so that after all the contours have been measured, a 'compromise' value of D can be chosen, which is not too far from the actual values for all the slabs. Then the extent E_i can be calculated for the boundary of

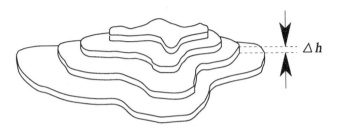

Fig 5.3 Method of measuring the surface dimension of an island from a contour map (see text).

any particular slab (labelled slab i), if n_i steps of length s were needed to step around it, thus:

$$E_i = n_i s^D. \tag{5.3}$$

The contribution of that slab to the extent of the surface is E_{si}, where:

$$E_{si} = hE_i. \tag{5.4}$$

E_{si} is obtained by multiplying a height (dimension 1) by a boundary extent (dimension D), so that the dimension of the surface is $1 + D$. If a pyramid were constructed from a stack of Koch islands of diminishing size, the dimension of the contour line at each level would be $1 \cdot 26$, and the dimension of the surface would therefore be $2 \cdot 26$. The extent of the entire surface is found by adding up the contributions for all the slabs:

$$E_s = \Sigma_i E_{si}. \tag{5.5}$$

This method is satisfactory provided that the contour dimension does not vary too much from level to level. If a particular contour is much smoother or rougher than assumed by the chosen value of D, then its contribution to the surface extent will not be independent of the step length used to measure it. Some real mountains and islands are about equally rugged all over, while others have flat or smoothly domed tops and rugged sides. If the contour dimension varies too much at different levels, then it may be necessary to divide the surface into different zones, within each of which the surface dimension is reasonably constant.

5.5 Density of organisms on fractal lines or surfaces

In some parts of their range, bald eagles, being fish-eaters, build their nests on coastal cliffs or stacks. A common-sense way of expressing the 'nest density' of a bird which does this is to give the number of nests per kilometre of coastline, and many such estimates can be found in the ornithological literature. Unfortunately, as explained above, typical coast-lines do not possess any such property as 'length'. The nearest usable property that they possess is fractal extent, suggesting that one should express nest densities per metron rather than per metre. One could indeed compare densities on two coastlines of the same dimension in this way, but if the dimensions of the coastlines differ, then so do the dimensions of the metrons used to measure them. This was the case in the comparison mentioned above between bald eagle nest densities around the coast of two islands (Amchitka and Adak) in the Aleutians. At the small map scale of Fig. 5.4, the coast of Amchitka appears rather smooth, but maps of larger scale show an abundance of offshore islets and stacks. These have to be taken into account in calculating the dimension, which has the rather

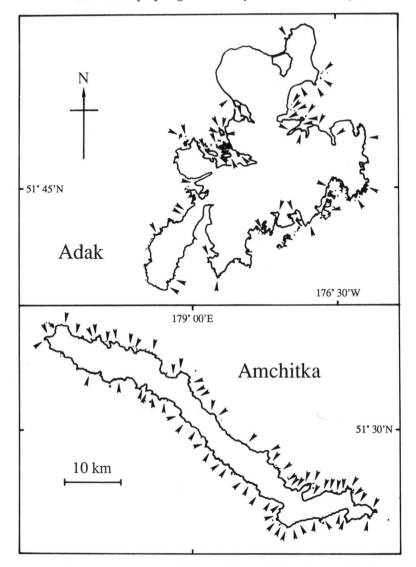

Fig 5.4 Bald eagle nests on Amchitka and Adak Islands in the Aleutians. The dimension of Amchitka's coastline (1·66) is much higher than Adak's (1·20) because the coast is broken into gullies and stacks on a scale too small to show on the map (from Pennycuick and Kline, 1986).

high value of 1·66. The coastline of Adak is more indented on a large scale, but there are not so many offshore islets, and its dimension is only 1·20. Amchitka had 66 nests, and the extent of its coastline was 1·51 × 10⁷ metrons of dimension 1·66, while Adak had 47 nests, and an extent of 1·08 × 10⁶ metrons of dimension 1·20. One cannot say that the Amchitka's extent is 14 times that of Adak, any more than one can compare an area directly with a length. The ratio of the extents is not a pure number, but has dimensions $L^{(1\cdot66-1\cdot20)}$. Before a direct comparison can be made, both extents must be reduced to the same dimension, so that the ratio of the two is a pure number. A simple solution to this is to use a number called the 'spacing' (S), defined as:

$$S = (E/N)^{1/D}, \qquad (5.6)$$

where E is the extent of the whole coast including the offshore islets, N is the number of nests, and D is the dimension. Raising an extent of dimension D to the power $1/D$ produces a length (dimension 1), just as raising an area to the power 1/2 (i.e. taking its square root) produces a length. The spacing so calculated works out to be 1·69 km for Amchitka and

A

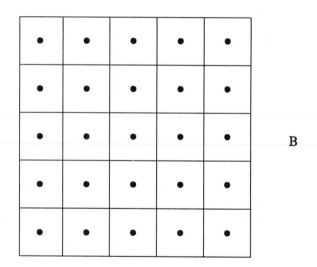

B

Fig 5.5   A, The 'spacing' of nests distributed along a road is simply the distance between them. B, For nests regularly distributed on a two-dimensional surface, the spacing is the area per nest, raised to the power 1/2, which is the same as the spacing in A. Spacings on fractal lines or surfaces are found in the same way, by raising the extent per nest to the power $1/D$.

4·31 km for Adak. The nest density is 2·6 times greater on Amchitka than on Adak, even when due account is taken of Amchitka's more rugged coastline. A similar method could be applied to comparing nest densities on two hillsides, with different dimensions between 2 and 3.

Although the spacing (*S*) is not something which can be measured with a tape measure, it is a simple length (dimension 1), which can be expressed in metres, and it does not depend on the step length used to measure the coastline. If the nests were distributed along a smooth line, as with osprey nests on power poles along a road, the 'spacing' would just be the average distance between nests. Again, if the nests are on a grid pattern on a smooth plain (dimension 2), then the spacing is found by taking the square root of the average area per nest. Once again it is the same as the distance between neighbouring nests (Fig. 5.5). Nest densities expressed as number of nests per metre or kilometre are valid when the distribution is on a smooth line such as a road, but are meaningless when used on coastlines, rivers, etc. Similarly, nest densities can be measured per unit area on a flat plain, but it is highly misleading to calculate a nest density by dividing the number of nests on a steep mountainside by the projected area of the mountain, as measured on the map. The 'spacing' works in any of these situations, and the reciprocal of the spacing ($1/S$) is a measure of density which gives directly comparable results on fractal surfaces of any dimension.

5.6 Fractals in morphology

Mandelbrot (1983) discussed a number of biological structures which have fractal properties. For example, a hierarchically branching system of vessels, such as a vertebrate arterial or venous plexus, looks much the same when viewed with the naked eye, as it does under a microscope. The pattern of branching is 'self-similar' over a wide range of scale. The surface area of the whole plexus, as measured by tiling with ever-smaller tiles, probably would eventually level off on the scale of the ultimate capillaries, but its value is poorly defined. If the branching continued to indefinitely small scales, then the surface area would be infinite, even though the whole plexus is contained in a finite volume. The notion of 'surface area', applied to a plexus, is a special case of fractal extent, with dimension exactly 2, but this particular type of surface does not correspond in a realistic way to the anatomical surface of a real plexus. To obtain meaningful and consistent measurements, the 'surface extent' would have to be measured, after determining the dimension (between 2 and 3) from the geometry.

It seems that absorptive surfaces, as in respiratory or digestive organs, are normally fractal in character. Mandelbrot cites apparently conflicting measurements of the alveolar 'area' of the human lung. If estimated from light microscopy, it comes out to be about 80 m^2, whereas an estimate from electron microscopy at higher resolution gives 140 m^2. Mandelbrot says that these, and other measurements made over a wide range of different scales, can be reconciled if the surface is not an area, but a fractal with dimension 2·17. This means that it is wrong to measure lung surface 'area' in square metres, and impossible to get a consistent result by so doing. In the terminology of this chapter, the 'extent' must be measured in metrons of dimension 2·17. In that case, the result will be unique, and will not depend on the scale of measurement.

It should be noted that real fractal structures never maintain self-similarity down to infinitely small scales, unlike mathematically constructed fractals, which can do so. Mandelbrot (1983) illustrates this with the example of a natural coastline, part of which has been modified by the construction of a straight sea wall. On a small-scale map, the sea wall would not show, and the coastline would have a dimension between 1 and 2 like any other. On a larger-scale map of the harbour area, however, the artificial sea wall would stand out as a section of the coast with dimension 1. Still closer inspection of the individual stones along the base of the sea wall would most probably reveal that their outline is once again a fractal, though not necessarily with the same dimension as the coastline as a whole. When dealing with biological structures or landscapes, a range of scales has to be identified, over which the structure is self-similar with a measurable dimension. The same structure may have different dimensions in different ranges of scale.

5.7 Fractals and allometry

If biological structures are liable to be fractals, then we have to rethink the principle of using allometric graphs to check for deviations from geometric similarity. Suppose you carry out some procedure which allegedly measures lung 'surface area' in a series of animals of different size, and make an allometric plot of surface area v. body length. According to the argument of Chapter 4, the plot should be a straight line with a slope of exactly 2, if the animals are geometrically similar. If the actual slope turns out to be greater than 2, then allometric reasoning leads to the conclusion that the animals are not geometrically similar. The larger ones must somehow differ in shape from the smaller ones, in such a way that their lung area is disproportionately increased. This is certainly a possible explanation for the observed result, but there is also an alternative possibility. Suppose, as Mandelbrot says, that lungs do not possess any such property as surface area. Their surface is not an area, but a fractal, with dimension 2·17. We should have plotted the extent, not the area (which does not exist). If we did that, then the slope would be 2·17, if the animals *are* geometrically similar.

The confusion about lung surface areas, which Mandelbrot found in the literature, was caused by applying a procedure suitable for measuring an extent of dimension 2 (that is, an area), to a structure whose dimension is actually $2 \cdot 17$. Physiological measurements are usually not applied directly to a geometrical structure, but to some process which is assumed to be related to the extent of the structure, whatever its dimension. An example of such a process would be the absorption of oxygen by a mammal, whose rate might reasonably be assumed to be proportional to the lung's surface area (if the lung had a surface area). If the lung does not have a surface area, but is a fractal with dimension $2 \cdot 17$, then one would expect the rate of absorption of oxygen in geometrically similar animals of different size to vary in the same way as the extent of the lung surface, that is, it would vary with the $2 \cdot 17$ power of the body length. If the rate of oxygen absorption were plotted against the mass rather than the length, as is usually done, then the slope would be $2 \cdot 17/3 = 0 \cdot 723$. As noted in Section 4.10, this is within the range of values commonly observed, when observations of basal metabolic rate are plotted against body mass, for a series of related animals.

5.8 Scaling of cruising locomotion

In his original analysis of the energy requirements for locomotion, Hill (1950) used the 'strength-limited' argument to deduce that stepping frequency should vary inversely with the length in geometrically similar, running animals. As the work done in each cycle is proportional to the mass of muscle, and thus to the body mass (or the length cubed), the mechanical power output would vary with the two-thirds power of the mass (or the length squared). The surface area of the lungs would also vary with the two-thirds power of the mass, and so the rate at which oxygen would need to be absorbed through each square metre of lung surface would be the same in animals of all sizes. It was noted in Section 4.7 that Hill's strength-limited argument is only appropriate for top-speed locomotion, which is normally anaerobic and not directly related to the rate of oxygen absorption. The stepping frequency in *cruising* running apparently varies with the $-1/2$ power of the length, rather than with the -1 power as Hill assumed, implying that the mechanical power output of the muscles would vary with the $5/6$ power of the mass in geometrically similar animals, rather than with the $2/3$ power.

An animal in prolonged, cruising locomotion must achieve oxygen balance, that is, its lungs must absorb oxygen at the same rate that the muscles (and the rest of the body) consume it. If the lung surface is an area (dimension 2), then its extent in geometrically similar animals would vary with the $2/3$ power of the mass, while the rate at which oxygen must be absorbed in cruising running would vary with the $5/6$ power of the mass. To maintain oxygen balance, the larger animals would have to absorb oxygen at a higher rate through each square metre of surface, or alternatively, the lung surface could scale allometrically, so as to vary with the 5/6 power of the mass. This would require the larger animals to have relatively bigger lungs than the smaller ones. The existence of fractals raises a third possibility. The animals might be geometrically similar after all, but with lungs whose surface is not an area, but a fractal of dimension D. Oxygen balance could be achieved if $D = (5/6) \times 3 = 2 \cdot 5$. This is higher than the value of $2 \cdot 17$, which Mandelbrot deduced for the dimension of the human lung, and would represent a highly convoluted surface, but then, people are not cursorial animals, adapted for prolonged cruising running. Possibly the lungs of antelopes, which are so adapted, have a higher dimension than human lungs. It is also possible that the extent of their lung surface might vary allometrically, and so might the rate of absorption per unit of surface extent, all at the same time.

5.9 Flying and swimming animals

Flapping frequency in flying animals also varies with the $-1/6$ power of the mass in cruising flight (other things being equal) as in running animals, so that the power available from the muscles in geometrically similar birds or bats varies with the 5/6 power of the mass, as in running animals. Bird lungs are required to supply oxygen at a rate which varies at least with the 5/6 power of the mass. They are very different in their anatomy and physiology from mammal lungs, and their internal structure is even more highly convoluted. This is usually explained on the basis that bird lungs can extract a greater percentage of the oxygen from the inspired air, because the flow through the actual lung appears to be unidirectional, from the posterior to the anterior air sacs. Also the ability to ventilate the air sacs independently of the lung may be important in controlling evaporative cooling in these well-insulated and highly active animals. However, there may be another reason for the unusual structure of bird lungs. It could be an adaptation to achieve a fractal surface with a high dimension, around $2 \cdot 5$, so that birds of different sizes can maintain oxygen balance in cruising flight, while maintaining geometrical similarity of the lungs. If the fractal dimension of bird lungs is higher than that of bats, it would follow that bats would require allometry of the lungs to a greater extent than birds (which might not require any allometry in their lungs). This might account for the puzzling fact that the largest flying birds, in several different orders, attain a body mass of about 12 kg, or possibly a little more, whereas the mass of the largest flying foxes is only about $1 \cdot 5$ kg. No doubt pterosaurs also had lungs with a high dimension, notwithstanding that the big Cretaceous pterodactyls may have flown in gravity that was weaker than it is now. The tracheal systems of insects, those age-old rivals of flying vertebrates, also have every appearance of being fractal surfaces of high dimension.

The use of fractal structures can permit a particular shape of animal to be scaled over a wide range of sizes without resorting to allometry. If no such expedient is possible, allometry becomes inevitable, that is, the body has to be distorted as the size is changed up or down. Eventually this distortion becomes so extreme that further increase (or reduction) in size becomes impractical. In muscle-powered flying animals, there is an unavoidable allometry of the wings, which severely limits the maximum size of birds and bats. The power *required* for cruising flight varies with the 7/6 power of the mass, whereas that *available* from the muscles varies with the 5/6 power of the mass, as above. The wings of birds are strongly allometric, in a way which helps to counteract this mismatch between power required and power available (Section 4.3). Even so, the largest flying birds, like condors and albatrosses, are small animals by the standards of mammals and reptiles, and they are only marginally capable of level flight by muscle power alone. Much larger flying animals would depend absolutely on an external source of energy (soaring), unless gravity were temporarily relaxed.

The situation is different in swimming animals, where the muscle contraction frequency is expected to vary with the –1/3 power of the mass rather than the –1/6 power (Section 2.6). This is the same relationship that Hill (1950) assumed, and so one might expect his conclusions to apply in the case of whales. The alveolar surfaces of their lungs would have to have smoother surfaces than those of cursorial mammals, so smooth that they could be said to possess a surface area (dimension 2). This is something that could probably be checked from published morphological descriptions. The gills of fishes would also be expected to possess an area, with dimension 2.

5.10 Feeding adaptations

One of the conclusions from Hill's (1950) dimensional analysis was that the surface area of a herbivorous mammal's molars would vary with the 2/3 power of the mass in geometrically similar animals, and so would the rate at which food could be ground up, thus matching the rate at which Hill deduced that energy is required for locomotion. The food is absorbed through the gut wall, whose area (if it possessed an area, as Hill assumed) would also vary with the 2/3 power of the mass. The rate at which energy is absorbed from food does not have to match the rate at which it is consumed so closely as oxygen supply must match demand in aerobic running. Animals may take in food at a rate in excess of consumption for prolonged periods, in which case they fatten up or reproduce, while if their food supply is not enough to cover their routine expenditure, they get thinner and eventually starve. It is widely believed that on average, the rates at which animals require energy for all purposes vary with the body mass in much the same way as the basal metabolic rate, that is roughly with the 3/4 power of the mass (rather than the 2/3 power as Hill assumed). This is rather a bold generalization, based on slender evidence, but I shall use it for want of a better one. Regardless of the exact value of the exponent, if the practical value is more than 2/3, then both guts and grinding teeth must either exhibit allometry, or have fractal surfaces, or a combination of both. After reading Mandelbrot's (1983) discussion of biological surfaces, it would be difficult to believe that the absorptive surfaces of intestines are *not* fractal in character. If their dimension is around 2·2 or higher, which seems not unreasonable, then herbivores of any size would be able to meet their nutritional needs without deviating from geometrical similarity of the guts.

Teeth are another matter. Although the grinding teeth of herbivores commonly exhibit a complicated pattern of ridges and cusps, the pattern is not self-similar over a range of scales. A tooth surface viewed through a lens or a microscope looks smooth, apart from scratches. It is not a fractal surface, because the large-scale pattern is not repeated at smaller scales. The tooth surface really does possess an area (dimension 2), and it follows that allometry would be needed in the grinding teeth of herbivores whose food requirements vary with the 3/4 power of the mass. This has not been investigated, but it seems very likely that large herbivores do indeed have a proportionately larger grinding surface, and also probably grind their teeth away at a greater rate. Small herbivorous mammals like voles and rabbits have a single row of relatively simple cheek teeth, which last the life of the animal. Larger herbivores like antelopes and horses have long snouts to accommodate a single row of large grinding teeth, which have a complex pattern of ridges, and sufficient length to permit prolonged wear. Rhinoceroses conform to the same pattern, but have proportionately very large heads and short necks, which limit the vertical range over which they can collect food. Elephants have huge grinding molars, and an unusual system of tooth replacement which greatly prolongs the functional life of the teeth. Elephants are also shaped differently from smaller herbivores, with a short, thick neck to support their enormous heads. The elephant's unique adaptation, its trunk, compensates for the lack of vertical mobility of its head, and allows it to collect food from ground level, or high up in trees, and pass it to the heavy grinding mill in its jaws, which stays at a constant level. There were larger herbivores in the past. Ornithischian dinosaurs, which were almost exclusively herbivorous, had jaws with differentiated teeth remarkably similar in general appearance to those of ungulates, although their morphology and methods of replacement were different. In particular, the hadrosaurs ('duck-billed dinosaurs') are famous for their huge jaws with multiple rows of grinding cheek teeth, which were replaced, when worn out, by new teeth erupting from below.

It would be difficult, but not impossible, to devise a method of measuring the area of the grinding surface in a herbivorous mammal or dinosaur with grinding cheek teeth. The surface clearly is an area, with dimension 2, and one would expect to see a stronger allometry than in the guts of the same animals (which might not show any allometry). Some groups of herbivores, such as antelopes and hadrosaurs, include animals of quite a wide range of different sizes, probably wide enough to look for allometry of the area of the grinding surface within the group. If this allometry exists, it might limit the maximum size of a herbivore with grinding teeth. Ever-larger herbivores would become progressively more top-heavy, as the size and weight of the head and jaws must increase out of proportion to the size of the rest of the body. This might explain why the largest ornithischians were only a little bigger than elephants, and not nearly as big as the largest dinosaurs of the other order, the Saurischia. The giant sauropods were the largest herbivores that ever lived, but they had relatively small heads with only plucking teeth at the front of the jaws, and no grinding cheek teeth. They apparently ground up their food in an enormous gizzard, housed in the body cavity. The gizzard stones have been identified in some sauropod fossils, lying in a pile where the body cavity used to be. The small, mobile head, on the end of a long, flexible neck, presumably gathered food in much the same manner as an elephant's trunk, and then passed it to the heavy, static grinding mill. By keeping the heavy machinery stationary in the centre of the body, this arrangement would permit the sauropod to reach a larger size, without affecting its mobility in gathering food.

Where allometry is unavoidable, as in bird wings, and the heads of herbivores that use teeth to grind their food, the resulting distortion of the body may set limits to the range of size that can be covered with the same basic body plan. Where fractal structures are involved, allometry may be avoidable, that is, geometrical similarity may be possible for some organs, especially respiratory and digestive organs, where it used to be thought that allometry must occur.

6. The functioning of ecosystems

6.1 Variables for describing ecosystems

There is a long tradition in theoretical ecology based on constructing differential equations that relate the number of individuals in a population and the reproductive and mortality rates of different age groups, with various external variables representing the physical environment and the status of populations of other species. However it is not possible, as the pioneers hoped, to predict the future course of population changes by methods of this type. They suffer from the same fundamental difficulty as weather forecasting. In the case of weather, the variables whose local values describe the condition of the air, such as its temperature, pressure, velocity, and moisture content, can be identified and measured at a large number of sample points, and used to construct a computer model of a portion of the atmosphere, whose behaviour can then be extrapolated into the future. In Britain, where information arrives several times per day from a dense network of observing stations, and much work has been done on perfecting the computer models, the accuracy and detail of the forecasts is impressive for periods up to three days ahead. Beyond that, the agreement between the model and the real atmosphere becomes rapidly less reliable. The trouble is only partly due to the incompleteness of the model and the limited amount of observed data. Even if the real atmosphere were strictly deterministic (like the model), and even if its condition were known to a high degree of precision at a given time, a very small change in the initial conditions (below any conceivable precision of observation), would lead to a somewhat larger change at a later time, which would lead in turn to a still larger change and so on. An initial change which is too small to measure can result, after only a few days, in a massive change in the entire system. Weather systems belong to a class of physical systems technically known as *chaotic*. A chaotic system is one which is deterministic, but whose future behaviour cannot be predicted indefinitely into the future, however precisely its initial conditions are determined.

Populations of organisms may or may not be deterministic, but even if they are, their development over time is not predictable far into the future. This is because changes in the initial conditions that are too small to observe or measure, are liable to 'snowball' into large effects in a relatively short time. The difficulty is compounded where many populations interact with each other, and with their physical environment, to form an ecosystem. Compared with the horrendous complexity of even the most rudimentary ecosystem, a weather system seems quite simple. Apart from the complexity of ecosystems, it is difficult to describe their functioning in terms of simple physical variables, operating on a small scale, comparable with the pressure, temperature, etc. of the air. Ecosystem processes seem to operate on an altogether different level from small-scale processes which, like the subject matter of Chapter 3, can be described in terms of simple variables like stress and strain. This has had the unfortunate effect that ecologists, while fond of intricate statistical analysis, are not always meticulous about identifying the dimensions of their variables. Fundamental errors, like confounding weight and mass, pass without comment in ecological journals, and have on occasion even been inserted by editors. Not even widely used variables like 'biomass' can be relied on to have fixed definitions or dimensions.

I have listed some variables which have been used for measuring the overall properties of ecosystems, with their dimensions, in Table 1.1 'Biomass' is the same as mass (dimensions M), the prefix 'bio-' simply indicating that it is the mass of living material. However, the same term is often used to refer to a different variable with dimensions ML^{-2}, which I call 'biomass density'. This is biomass per unit area (not per unit volume like ordinary density) because the amount of biomass that can exist in an ecosystem depends on the area over which it receives radiation, not on the volume of soil or water that it occupies. 'Production' is used in two different ways by ecologists, as the rate of generation of either biomass or energy (dimensions MT^{-1} or $ML^2 T^{-3}$). Since the biomass created by photosynthesis goes hand in hand with the Gibbs free energy incorporated into it, these two measures of production are roughly equivalent. Ecologists distinguish 'primary' and 'secondary' production as the rates at which plant and animal material is generated, respectively. 'Productivity' is production per unit surface area (dimensions $ML^{-2} T^{-1}$ or MT^{-3}). The problem of measuring production on a fractal surface has not yet been addressed.

Variables with the dimensions of mass (though not necessarily identical with total biomass) and rate of flow of mass, can be used to relate the properties of individual organisms, as discussed in earlier chapters, to the functioning of populations, and of ecosystems. Of particular importance is a variable formed by dividing a mass flow (dimensions MT^{-1}) by a mass (dimensions M) to yield a pure rate (dimensions T^{-1}). I do not think that this variable has been identified before, and I shall suggest that it is the key to relating the properties of an ecosystem to those of the organisms of which it is composed. First, a schematic framework is needed, to show how masses and mass flows are organized.

6.2 A schematic ecosystem circulation

The general principles governing the flow of matter and energy in ecosystems have been understood for many years. The purpose of constructing

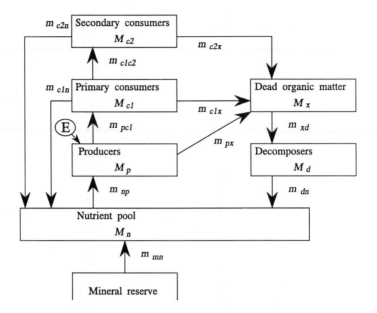

Fig 6.1 Flow diagram for a simplified natural ecosystem. The capital M symbols represent the mass (dimensions **M**) of a specified element in each of the closed boxes, which themselves represent ecosystem stages. The sum of the M's is the 'circulating stock'. The lower-case m symbols represent rates of flow of mass (dimensions $\mathbf{MT^{-1}}$) of the same element between stages, identified by the subscripts. The circulation is driven by solar energy, captured by the producer stage ('E'). The 'circulation rate' for the whole system is the same as the rate of flow from the nutrient pool to the producers (m_{np}). The 'mineral reserve' (open-sided box) is not part of the circulation. The net rate of flow from it to the nutrient pool (m_{mn}) represents one-time transfers of material.

a diagram such as Fig. 6.1 is not to discover anything new about this, but to simplify what is already known in a manner that allows explicit statements to be made about it. Figure 6.1 is not supposed to represent the infinitely complicated flows that occur in any real ecosystem, but rather the minimum number of distinguishable stages (or almost the minimum) which have to be present in an ecosystem that works. Actually it is possible to have an ecosystem consisting only of plants and a nutrient pool, but this is unexciting for zoologists. Two levels of consumers have been included, to allow some interactions to develop which can be seen in real ecosystems. First, the dimensions of the symbols in Fig. 6.1 have to be defined, and then we have to identify exactly what is circulating or flowing.

Each box in Fig. 6.1 is an ecosystem 'stage'. A stage is like a trophic level, but broader in definition, so as to allow for non-living components of the ecosystem. The stages are designed to account for *all* of the material that is circulating around the ecosystem, so there have to be stages for dead organic matter and the nutrient pool, as well as for the living trophic levels. The 'mineral reserve' is not an ecosystem stage, and an open-sided box is used to indicate this. Each closed box in Fig. 6.1 contains a symbol M, subscripted to identify the ecosystem stage to which it refers, thus: M_p for the producer stage, M_{c1} for the primary consumers and so on. These M symbols represent *the total mass of a specified element*, currently contained in that stage (dimensions **M**). The arrows connecting the boxes are identified by lower-case m symbols, subscripted to indicate the source and destination of each arrow. They represent *the net rate of flow of the same element* in the direction of the arrow (dimensions $\mathbf{MT^{-1}}$). If there is a net flow in the opposite direction to that shown, then the value of that m is negative. The idea behind Fig. 6.1 is that by including the nutrient pool and dead organic matter as ecosystem stages, the sum of the M's represents all of the mass of the chosen element in circulation in the ecosystem. This amount of material (ΣM) is called the ecosystem's 'circulating stock' of the specified element. Its rate of increase is equal to m_{mn}, which is the net rate of flow from the 'mineral reserve' (i.e. all those parts of the earth's crust which are not initially accessible as plant nutrients) to the nutrient pool. m_{mn} may be either positive or negative. New nutrients dissolved from bedrock, or imported in alluvial sediment, make a positive contribution to m_{mn}, while removal of material by soil erosion or otherwise, out of reach of the ecosystem's plants, makes a negative contribution. m_{mn} represents one-time transfers of material into or out of the ecosystem, and is not part of the circulation of material.

The M's of Fig. 6.1 have the same dimensions (**M**) as the familiar 'biomass', and the m's have the same dimensions ($\mathbf{MT^{-1}}$) as one version of the ecological variable 'production', but there is an important difference. Whereas 'biomass' is usually understood to refer to the total mass of living material, and production to its rate of generation, the masses and mass flows in Fig. 6.1 refer to *one element*. A complete, separate diagram, with its own set of M's and m's is required for each different element circulating in an ecosystem. The mass represented by each M of Fig. 6.1 is not the whole of the biomass in its box, but only that part of it consisting of the specified element.

6.3 Conservation of mass within a boundary

Treating the circulating stock as a conserved quantity implies that the ecosystem has a boundary, across which material does not flow. This is strictly true only if the 'ecosystem' is identified as being the whole of the biosphere. Lesser ecosystems can also be defined, which vary in the degree to which material flows respect an identifiable boundary. Many lakes can be considered bounded ecosystems, in which m_{mn} could be measured by observing stream inflows and outflows. Oceanic ecosystems mostly have to be inconveniently large to be considered bounded, but many terrestrial ecosystems of moderate size can be regarded as bounded, to a sufficient degree of approximation. An example is the Serengeti ecosystem of East Africa (Sinclair and Norton-Griffiths 1979), where the primary consumer stage is dominated by large herbivorous mammals, including a large population of peripatetic wildebeest. Practical ecologists in that area regard the 'Serengeti Ecosystem' as anywhere that the wildebeest go. A well-defined boundary can be marked on the map on that basis, and it probably is true that the flow of material across this boundary is small, compared with the circulating flow that goes on within the boundary. In that case, and in many other terrestrial ecosystems, Fig. 6.1 is a useful approximation to the real situation, but only for elements that do not circulate through the atmosphere. In the case of oxygen, for example, the nutrient pool is effectively infinite, and therefore so is ΣM. Except when the 'ecosystem' is the whole biosphere, the concept behind Fig. 6.1 is most realistic for elements like phosphorus or calcium, which do not circulate through the atmosphere. Many plant communities can be considered to have a circulating stock of nitrogen, whose total amount varies only slowly as a result of the activities of free-living nitrogen-fixing and denitrifying bacteria in the soil. These slow flows can be represented by m_{mn}, that is, as one-time transfers from the normally inaccessible mineral reserve, which in the case of nitrogen is the atmosphere. Leguminous plants, however, are able to exploit nitrogen-fixing bacteria to a degree which makes atmospheric nitrogen accessible to them as an effectively infinite nutrient pool, and in that case the idea of a bounded ecosystem only applies to the biosphere as a whole.

6.4 Energy flow and the circulation rate

Figure 6.1 does not represent the flow of energy. Energy enters an ecosystem via photosynthesis, and may flow through one or more stages before being lost from the ecosystem. It does not recirculate to any significant extent, and there is no question of a fixed 'stock' of energy being conserved. The 'E' sign in a circle, with an arrow to the producer stage, indicates that energy enters the system at that point, and drives the circulation of material. The flow of material entering the producer stage has a special significance. All other flows eventually lead to the nutrient pool, and the only route by which material can get back from there, into the living stages, is via the producers. m_{np} is therefore singled out and given a special name the 'circulation rate'. It is the *mass inflow* of one specified element into the producer stage. This is not quite the same as 'primary production', which usually refers to the rate of generation of plant biomass (all elements). However, the circulation rate for any given element may be assumed to be proportional to primary production, as usually defined. m_{np} is driven by photosynthesis, and is ultimately limited by the rate at which energy is supplied to the plants by solar radiation, although this limit is seldom approached in real ecosystems. Usually, m_{np} is limited to some much lower value by other physical factors, such as the availability of water for transpiration in terrestrial ecosystems, or nutrient availability in aquatic ones.

6.5 The mass accumulation principle

An important and universal characteristic of ecosystems, as represented in the manner of Fig. 6.1, follows directly from Darwin's principle of natural selection. In the terminology of Chapter 4, natural selection ensures that every organism is adapted to maximize its own power surplus. In biomass terms, every organism strives to convert as much material as possible into the biomass of its own offspring, and to prevent other organisms from purloining its own biomass until this has been achieved. Plants strive to maximize the rate at which they convert inorganic nutrients into their own biomass, and to minimize the rate at which biomass is lost to herbivores, while animals strive to maximize the rate of increase of their own biomass, at the expense of that of plants or other animals. No organism is concerned that a limited stock of available material has to be shared among different ecosystem stages, in order to maintain the system of which it is a part in working condition. The distribution of biomass is the result of automatic processes, not of the adaptations of individual organisms, which invariably endeavour to concentrate as much material as possible into themselves.

The mass accumulation principle applies to species populations as well as to individuals. Each of the 'living' boxes in Fig. 6.1 represents a set of species populations, which together constitute an ecosystem stage. The combined efforts of all the individuals of all the species in each box are directed towards maximizing the rate of flow of material into that box, and minimizing the rate of flow out of it. The organisms comprising any living stage of an ecosystem continually strive to concentrate *all* of the circulating stock of material into their own stage. Success in this endeavour would, of course, disrupt the ecosystem. Normally, the efforts of the organisms in any particular stage to monopolize the circulating stock are frustrated before too great an imbalance develops, by the counter-efforts of organisms in other stages. If these counter-efforts become ineffective for any reason, then disruption can and does occur.

6.6 Productivity alone does not determine biomass density

The fundamental limitation noted above acts on m_{np}, the *rate* at which material flows from the nutrient pool into the plants. This does not of itself limit the amount of the circulating stock. Even if the plants only withdraw material from the nutrient pool at a very slow rate, they could, given enough time, accumulate any amount of living material. However, it is obvious that something does limit the amount of material in the living stages at any one time. Biomass does not accumulate indefinitely. The average density of (dry) biomass over the whole of the earth's surface is only about $1 \cdot 2 \, \text{kg m}^{-2}$. If evenly spread over the whole earth, life would form a very thin scum, a fraction of a millimetre thick. Even over the continents, where most of the biomass is concentrated, the average density is only about $4 \, \text{kg m}^{-2}$. Biomass densities up to $40 \, \text{kg m}^{-2}$ have been recorded in forests, but other types of biomes support less biomass, typically $1-5 \, \text{kg m}^{-2}$ in tropical savannahs, and around $0 \cdot 5 \, \text{kg m}^{-2}$ for tundra. Nearly all of the biomass in most terrestrial ecosystems consists of plant material. Even in ecosystems with conspicuous populations of large herbivores, like the savannahs of East Africa, the animal biomass density is three orders of magnitude below that of the plants. Where smaller herbivores like grasshoppers predominate, the animal biomass density is less (Section 6.7).

Evidently some environments are more 'favourable' in some sense than others for the accumulation of biomass. The physical requirements for a 'favourable' environment are familiar enough — plenty of light, plenty of water and nutrients, and temperatures that stay well above freezing, but never approach a lethal level. There is also no mystery about the way in which these requirements translate into favourable ecological conditions. They promote high plant productivity, that is, a high rate at which plant matter is produced, per unit area of land or water surface. However, this is not the same thing as supporting a high biomass density (mass per unit area). Estimates of the productivity of the short-grass plains of the Serengeti in wet weather are not very different from estimates from forest biomes during their season of maximum production, but the biomass density is less on the short-grass plains by a factor of 30–50. The green *Spirulina* soup of the alkaline Rift Valley lakes has still less biomass density, but again its productivity is similar. In the forest, the new plant material accumulates to form the bodies of trees, whereas on the short-grass plains it is eaten by herbivores almost as soon as it is formed, and is passed along to the consumer stage. As for the small-bodied *Spirulina*, once they accumulate to a layer a metre or so thick, they cut off the light to their fellows below, which die and recirculate their nutrients to the layer above, if they are not eaten by animals first.

6.7 Intake rate — a new variable with dimensions T^{-1}

Productivity, or the related variable mass inflow (*m*), is predictable to some degree from the physical conditions, but biomass density, or total biomass (*M*), does not follow, because the dimensions are different. A variable is needed that connects the rate of flow into a box of Fig. 6.1, with the amount of material in that box. I shall call this variable the *intake rate* (*R*), where:

$$R = m_i / M. \qquad (6.1)$$

Intake rate must not be confused with *mass inflow*, denoted by m_i, which is the rate at which material flows into a specified box of Fig. 6.1. *M* is the amount of material in the box. *R* is obtained by dividing a mass rate (dimensions MT^{-1}) by a mass (dimensions **M**), so its own dimensions are T^{-1}. *R* is a pure rate, with the same dimensions as strain rate or frequency. The SI units are 'per second' (s^{-1}). The intake rate for an ecosystem stage is simply the sum of m_i for all the organisms making up that stage, divided by their total mass.

The notion of an intake rate can be applied to an individual organism as well as to an ecosystem stage, remembering that the *m*'s and *M*'s were defined above as referring to a particular element. For example, we could define an intake rate for sulphur for an individual animal (a goat, say). m_i in this case is the difference between the mass rate ($kg\,s^{-1}$) at which sulphur is consumed by the goat in its food, and the mass rate at which sulphur is lost in faeces, urine, moulted hair, abraded horns and hooves, etc. *M* (kg) is the mass of sulphur actually in the goat. The goat's intake rate (*R*) for sulphur is then m_i/M and is measured in units of s^{-1}. There are lower and upper limits to the possible values of *R*, which are physiological properties of the goat. The lower limit (R_{min}) expresses the minimum rate at which the goat needs to take in sulphur in its food, to replace unavoidable losses. R_{min} is the minimum intake rate, at which the goat just maintains its body mass. There is also an upper limit to the rate at which even the most voracious goat can chew, digest, and absorb food, and this will set a practical value for R_{max} for sulphur. R_{min} and R_{max} could be measured and tabulated for each element, and for different kinds of animals. This has not been done, but if it were, one can be confident that scaling relationships would appear, of the kind discussed in Chapter 4. Both R_{min} and R_{max} must be lower for large animals than for small ones. This is because the rate of flow of material through an organism is closely related to the rate of flow of energy. As noted in Chapter 4, allometric comparisons between animals of different size, but otherwise similar type, show that the mass-specific rates at which they process energy for any specified purpose typically vary with the –0·25 power of the body mass. I would go so far as to predict that when comparing similarly adapted and related animals (herbivorous mammals, say), the ratio of R_{min} to R_{max} would not vary much, and each of them would vary with a power of the body mass near to $-0 \cdot 25$. At any rate, no zoo-keeper would dispute that far less food is needed, per day, to feed an elephant of given mass, than to feed a large collection of mice with the same total mass. Andean peasants prefer to raise guinea-pigs rather than llamas as a source of meat, because a smaller mass of guinea-pigs than of llamas is needed to generate meat at a given rate, and they take up less space. I shall argue later in this chapter that if empirical data on the scaling of R_{min} and R_{max} did exist, they could be used to predict the response of an ecosystem to disturbances.

6.8 Scaling of intake rate of plants

The mass inflow (m_i) of some element into a plant, or a population of plants, is the rate at which that element is taken up in the form of low-energy plant nutrients, to be converted into higher-energy compounds by photosynthesis. This is not quite the same thing as 'primary production', which is the rate at which plant dry matter (all elements) is produced, but it resembles primary production in being directly linked to the rate at which energy is captured in photosynthesis. In production terms, about 40 MJ of captured energy corresponds to each kilogram of carbon fixed. In terms of intake rate, each kilogram of any specified element, absorbed in the course of photosynthesis, requires a definite amount of energy, the actual amount depending on which element it is. Elements other than carbon, hydrogen, and oxygen are not directly involved in photosynthesis, and are taken up from the transpiration stream in land plants, or directly absorbed from the water in aquatic ones. However, the proportions of other elements, such as sulphur, phosphorus, and nitrogen in plant matter as a whole are only moderately variable, and the amounts of these elements taken up by plants also correspond approximately with the amount of energy captured.

The mass inflow of a particular element for plants on a given area of the earth's surface (m_{np}) is ultimately limited by the rate at which solar energy is supplied. The practical limit is well below this, further below for some kinds of plants than for others, depending on their special adaptations to acquire water or particular nutrients. Sunlight beats down impartially on each square metre of the earth's surface, regardless of what type of plants (if any) are spread out below to absorb it. Provided that plants of one kind or another are available to take full advantage of the opportunities supplied by the physical environment, productivity depends more on physical factors than on the size or other characteristics of the plants. If the available rate of mass inflow (m_{np}) is taken up by large plants (trees), M_p is large, and R_p is small. If the plants are smaller (algae), M_p is smaller, but m_{np} (being determined mainly by the physical environment) is much the same as before, so R_p is larger. Thus the intake rate (R) for a plant, or a population of plants, must be a decreasing function of body size (or mass), in much the same way as for animals.

The notion of 'body mass' is one that comes more naturally to a zoologist than to a botanist. It is a simple job to weigh an animal, so long as it is not too big, but weighing a tree is much harder, because of the difficulty of digging up the roots. Besides, most of the wood in the interior of a tree trunk is actually dead. Should a tree's 'body mass' include the dead wood, or only the living tissues, which would be a small fraction of the mass of the whole tree? One could ask the same question about a snail's shell, an antelope's horns or a bird's feathers. For the ecological argument which follows, the answer is that such dead structures should be included as part of an organism's body mass, if they were created by the organism's own metabolic efforts, and if they continue to form an integral part of its structure. The deer's antlers count as part of its biomass until it sheds them, and at that moment the material they contain passes from the 'primary consumer' box to the 'dead organic matter' box. As a tree grows, its mass inflow (m_i) increases as the root system spreads out, but its body mass (including wood) increases more, and the intake rate (R) therefore declines.

Land plants in general tend to be much more variable in size than most animals, and consequently 'body mass' is not usually seen as a good variable for characterizing plants. Variable they may be, but different types of plants do consistently occupy different size ranges. Plant species which, when fully grown, normally form part of the rain forest canopy definitely have larger bodies than unicellular algae, even though there may be some difficulty in defining exactly what constitutes the plant's 'body'. Shrubs and herbs occupy intermediate size ranges. If one concedes that there is some meaning, albeit not a very exactly definable one, in speaking of the 'size' or 'mass' of an individual plant, then R_{min} and R_{max} must depend on body size in plants as well as animals. As in animals, these minimum and maximum intake rates are physiological characteristics of different species, which could be measured and tabulated. Other characteristics in addition to body size must contribute to determining R_{min} and R_{max} in different species at different stages of growth, but it is fair to assume that there is a strong overall trend for these values to be lower in large plants than in small ones. The same conclusion would apply if R were measured for the producer stage as a whole, rather than for individual plants. If the mass inflow to the stage (m_{np}) were divided by the biomass in the stage (M_p), the result would be higher for a grassland than for a woodland, given similar environmental conditions.

6.9 Maximum biomass or carrying capacity

For an animal population, there is an upper limit on the mass inflow (m_i), which in informal terms can be equated with the available food supply. If the maximum available value of m_i is known, then the maximum biomass of that species that can be supported by the given food supply is determined by R_{min}, since

$$M_{max} = m_i/R_{min}. \tag{6.2}$$

It is a reasonable presumption that R_{min} for an animal, or a population of animals, is a decreasing function of body mass, varying with the -0.25 power of the mass, or something near to that. Since m_i represents an externally determined food supply, which is not affected by body mass, M_{max} may be expected to vary with the $+0.25$ power of the body mass. Suppose we have a fixed mass of food per day, which is to be used to support either elephants (mass 2000 kg) or mice (mass 20 g), then the elephant:mouse ratio for M_{max} will be $(10^5)^{0.25} = 17.8$. About 18 times as much biomass of elephants can be supported on a given food supply as if the same, fixed food supply were fed to mice. The rule may not be exactly the same for plants, but it certainly is true that a given rate of mass inflow from the nutrient pool supports a larger biomass of large plants than of small ones.

'Carrying capacity' is usually defined as sustainable biomass per unit area, whereas M is the total mass of a single element in a species population or an ecosystem stage. However, carrying capacity, when expressed in terms of biomass density, is higher for large animals than for small ones, and for large plants than for small ones, in the same way as M.

6.10 'Habitat destruction' by herbivores

A deeply ingrained belief among wildlife managers is that herbivore populations 'need' to be managed by human managers, otherwise they are liable to 'destroy their own habitats'. The idea is that if herbivore numbers are allowed to increase unchecked, then the herbivores will eat the vegetation down to a point from which it cannot recover, whereupon the herbivores starve *en masse*, and the habitat degenerates irreversibly into unproductive desert. Perhaps it happened to the dinosaurs—who knows? One of the most widely publicized instances of 'habitat destruction' occurred in Kenya around 1959-60. Following the establishment of Tsavo National Park in the early 1950s, elephants were protected from ivory poachers, whose depredations had held the population density down to a level far below any probable carrying capacity, for the previous half century or more. As far back as anyone could remember in 1959, the landscape had been mostly covered by mature woodland, dominated by various species of

Commiphora. From the air, this is a somewhat monotonous vegetation type, with red lateritic soil visible through and between the regularly spaced trees, and sparse populations of large herbivores, at densities far lower than those that prevail in the grasslands of the Serengeti. As their density built up, the elephants began to show an increasing propensity to push over *Commiphora* trees, as a quick and easy way to browse the branches at the top. In 1959 the rains failed, and plant production, which corresponds closely to rainfall in that area, all but ceased. As the long drought progressed, the elephants, with little else to eat, gradually reduced a large area of the eastern part of the park to a terrible dust-bowl strewn with the shattered skeletons of *Commiphora* trees. Many elephants starved. They had destroyed their habitat — or so it seemed at the time.

Eventually the rains resumed, and a run of good rainfall years followed. Ten years after the drought, Tsavo East was not a desert, but neither had it reverted to the pre-drought status quo. It was accurate to say, as local ecologists did, that the *Commiphora* woodland had been 'destroyed', but the implication that the ecosystem had been destroyed, or severely damaged, was more questionable. With the next rains, fast-growing shrubs and herbs took over the area that had been largely cleared of *Commiphora* trees. Patches of grassland developed, which burned in later dry seasons, so inhibiting tree regeneration and maintaining the grassland. Grazing animals such as zebra and hartebeest, which had previously been scarce in the area, moved in to exploit the more open habitat. The new plant community was radically different from the earlier *Commiphora* association, but was also more varied, and appeared to be as productive as the physical habitat would allow. The ecosystem continued to function, but with a different plant association, dominated by annuals and fast-growing herbs and shrubs, instead of the long-lived *Commiphora* trees.

It is understandable that managers of national parks become alarmed when they see a familiar stand of trees getting decimated by elephants. Actually, equally loud complaints are heard in other parts of Africa to the effect that 'overgrazing' by herbivores inhibits grass fires, and causes grassland to 'degenerate into bush'. Landscapes, including natural vegetation, are seen as elements of our environment which are expected to be stable and permanent. On a time-scale of a few thousand years, however, dramatic changes of vegetation cover are the rule rather than the exception. Moreau (1966) gives a fascinating account of the way that montane forests spread over the plains of Africa during the ice ages, but retreated to the mountains during interglacials, to be replaced by more open savannah vegetation at lower altitudes. Areas of woodland also come and go on a time-scale not much longer than a human lifetime. As a case in point, accounts by early European explorers suggest that the area that is now Tsavo National Park was open savannah in the mid-nineteenth century. It is quite possible that the *Commiphora* woodlands found up to the

1950s developed in response to the decimation of elephants by ivory hunters, which took place in the 1880s. It could be argued that the opening up of the habitat which followed the 1959–60 drought was not 'habitat destruction', but a reversion to an earlier, more open vegetation type, in response to a more 'normal' density of elephants.

6.11 Response of plants and herbivores to drought

It is instructive to think about the effect of a severe drought on an African savannah ecosystem, in terms of the circulation of a fixed stock of material, as in Fig. 6.1. To understand the response of savannah plants to a drought, it is important to be clear as to what plants are adapted to do. Some people, notably those with an agricultural bent, imagine that plants are adapted for animals to eat, but that would be contrary to natural selection. Plants are adapted to convert nutrients into their own biomass, and also to prevent that biomass from being diverted into the bodies of other organisms, i.e. herbivores. In the symbolism of Fig. 6.1, plants are adapted to maximize m_{np}, and to minimize m_{pcl} and m_{px}, so accumulating the maximum possible amount of material into M_p. Providing food for animals is a hazard for plants, which they avoid or minimize by all means at their disposal. An actively growing and photosynthesizing plant has no alternative but to put out new leaves, which are vulnerable to attack by herbivores. Once a spell of dry weather sets in, the supply of water for transpiration quickly ceases, and there is then no longer any need to display edible tissues for the attention of herbivores. Savannah plants respond to dry weather by translocating the more edible components of their leaves underground into the root system. The leaves turn brown, and only the structural elements, mostly cellulose and lignin, remain above ground. This condition is reversible in many savannah grasses and herbs. In the event of rain, the apparently 'dead' leaves turn green again and resume photosynthesis. The energy expenditure required to dig up, eat, and digest material that has been translocated into plant roots is evidently great enough to discourage most savannah herbivores from attacking the roots, presumably because the anticipated income is too meagre to yield a positive power surplus. Translocation of material underground is an effective method of defending plant material against herbivores, during periods when the plants cannot photosynthesize, because of lack of water for transpiration. The effect of drought on savannah vegetation is that m_{np} drops nearly to zero as the soil dries out, and transpiration ceases. The plants respond by turning off m_{pcl}, the onward flow to the herbivores.

With the exception of a few fortunate holders of territories in areas supplied with ground water, most savannah herbivores find themselves making a negative power surplus, soon after the onset of a drought. This quickly translates into loss of body mass. Undernourished animals first use up reserves of fat, if they have any, then become emaciated as they are forced to consume muscle and other tissues. M_{cl} declines because of the loss of body mass of individual animals, long before animals actually start to starve and die. Because of the scaling relationships discussed in Chapter 4, larger animals take longer than smaller ones to starve to death in a drought, assuming that all begin in similar condition. When the plants turn off m_{pcl}, the largest animals last the longest. A habitat subject to droughts of uncertain duration therefore always provides selection pressure favouring large body size, but there are also other selection pressures affecting body size, in both animals and plants.

6.12 Automatic response of body size to material abundance

In the simplified ecosystem of Fig. 6.1, containing organisms obeying Equation 6.1, the mass (M) of some specified element in any particular box can be represented in terms of only two variables. In the case of the producer stage:

$$M_p = m_{np}/R_p. \qquad (6.3)$$

As noted in Section 6.8, m_{np} is primarily dependent on the physical characteristics of the environment, not on the characteristics of the particular species of plants making up the producer stage. M_p, the amount of material allowed to accumulate in living plant tissue, depends on m_{pcl} as well as on m_{np}. In the absence of herbivores, m_{pcl} would be zero, but if herbivores are present, they start removing material from the producer stage as soon as the plants produce it, so preventing M_p from building up. Heavy herbivore pressure holds M_p down, but usually not to such a low level that m_{np} is depressed in turn, below the level that the physical environment would permit. What happens is that the producer stage adjusts the value of R_p so as to reconcile the value of m_{np}, determined by environmental factors, with the amount of material available for plant biomass. If herbivore pressure is heavy, as on the short-grass plains of the Serengeti, the only plant species that can survive are small grasses and herbs which can grow and set seed in the short time available before herbivores devour their aerial parts. The herbivore pressure does not depress m_{np} (or plant production). Instead, it favours small species, with a high intake rate (R_p). If the herbivore pressure is relaxed, naturally or artificially, then the balance of advantage shifts to larger plant species. Experimental exclosures — plots from which herbivores are excluded by a fence — quickly change over to long-grass associations, dominated by grass and herb species which are normally characteristic of wetter areas. The biomass density increases dramatically, because the availability of more material causes the species composition to shift in favour of species with larger bodies and lower R_p.

In places where m_{np} is higher because of higher rainfall, and herbivore pressure is lower, woodland or forest can develop.

The same principle applies at the next level. The rate at which material enters the primary consumer stage, m_{pc1}, is determined primarily by the rate at which plants produce accessible, edible parts for herbivores to eat. This 'food supply' is fixed by constraints external to the primary consumer stage. The amount of material which can accumulate in the form of primary consumer biomass is determined by the intake rate for the particular species present:

$$M_{c1} = m_{pc1}/R_{c1}. \qquad (6.4)$$

If a generous supply of food is available, and the loss of material by herbivores due to predation and other forms of mortality (m_{c1c2} and m_{c1x}) is low, then material can accumulate in the herbivore stage, and populations of large animals (low R) can build up. If these are decimated, for example by shooting or epidemic disease, M_{c1} is reduced but m_{pc1} is not much affected. The slack is taken up by rodents, grasshoppers, and termites, which quickly multiply to take advantage of the available plant food. The consumer stage responds by redeploying a reduced amount of available material in smaller organisms, so raising R_{c1} and restoring the balance required by Equation 6.4.

6.13 Species shift versus ecosystem disruption

The type of self-regulation discussed in the last section can produce drastic changes in the appearance of plant and animal communities, with little or no change in the actual list of species present, or of the amount of circulating material. For example, the Serengeti ecosystem contains a number of even-aged stands of mature trees, which are obviously not in a steady state, because there are no younger trees or seedlings among them. The origin of some of these can be traced to past episodes of exceptionally heavy grazing pressure caused by the build-up of herbivore populations during a run of wet years. When a run of dry years then sets in, any available grass is heavily grazed, with the result that grass fires are suppressed in the dry season, because there is not enough uneaten material left to burn. The temporary cessation of dry-season fires permits a generation of tree seedlings to grow to a height at which they are able to survive fires. Herbivore mortality rises because of shortage of food, and populations decline. Surplus grass becomes available again to sustain fires, and further growth of tree seedlings is prevented — but a cohort of trees has established itself in an area that used to be open grassland. Once they are big enough to tolerate fires, the trees grow to maturity, and an area of woodland develops in what used to be grassland. However, there is no continuing regeneration of young trees, because the growth of new seedlings is suppressed by the regular dry-season fires. Eventually the trees become senile, and succumb one by one to assaults by elephants or wood-boring insects, or to changes in the level of the water table or other hazards. Gradually, the area reverts to grassland once again.

The rate at which these changes from grassland to woodland and back again take place is slow enough that managers can form the impression that the trees (or grassland) have 'always been there', and thus attribute conspicuous changes to 'habitat destruction'. The crucial distinction is whether or not the changes have depressed the circulation rate (m_{np}) below the level that the physical environment would otherwise allow. If plant production continues at much the same level as before, but with a different assemblage of plants, then it is more appropriate to describe the change as a 'species shift', whereby the ecosystem automatically adjusts the intake rates (R) of one or more stages, compensating for increased or decreased availability of material.

This type of self-regulation process depends for its operation on a sufficient diversity of species. The visual appearance of the more homogeneous vegetation types may be dominated by a few abundant species, but the diligent botanist can invariably compile a lengthy list of other species, occurring at low densities or as scattered individuals. A 'species shift' occurs when one or more species, that were previously present but uncommon, increase in density until they dominate the scene, while the previously dominant species dwindle to become rare. If the supply of material for plant biomass is reduced, then the system will respond automatically by shifting in this manner towards smaller species, and vice versa. The current high rate of extinction of species is a matter of extreme concern in conservation circles, and one of the reasons for this is that loss of species reduces an ecosystem's resilience, that is, its ability to compensate for forced mass redistribution by species shift.

What determines whether a particular species can thrive in given conditions is not body size as such, but the range of possible values of R which it can tolerate. For example if some nutrient (phosphorus, say) is in short supply, this generally favours smaller organisms, but certain kinds of trees may still be able to compete, by exploiting symbiotic associations with mycorrhizal fungi in their roots. R_{min} and R_{max}, whether determined by body size or special adaptations, are characteristics of different organisms which, if their values were known, would provide a basis for predicting which species would be likely to prosper and which to decline in response to given ecosystem disturbances. It would be a laborious but straightforward job to measure and tabulate R_{min} and R_{max} for different plants and animals at different stages of their life cycles. It is an effort which should be undertaken in some selected ecosystem, to test the predictive value of this approach.

6.14 Disruption by loss of circulating stock

A common way in which genuine, irreversible disruption of ecosystems occurs is by massive removal of circulating material, for instance by accelerated soil erosion, whereby material is withdrawn from the nutrient pool and removed from circulation. Actually, the effect is much the same if material is removed from any ecosystem stage. Removal of plant crops for market agriculture, or 'cropping' of wild or domestic animal populations also depletes the circulating stock, if material is taken away from the ecosystem and not replaced. For example, large areas of the Scottish Highlands, once forested, were progressively depleted of calcium by raising and removing sheep, and have now changed irreversibly in a way that makes the landscape incapable of developing anything but very poor grassland.

The contentious question of whether an ecosystem has been 'damaged' is perhaps best approached in terms of the physical environment. In terrestrial ecosystems, the prevailing combination of water availability, temperature, day-length, etc., defines a maximum possible value for the circulation rate (m_{np}) for any element, given the plant associations native to the area. It should be possible to devise a method for estimating the maximum sustainable value of m_{np} from physical measurements. Then, an ecosystem which no longer contains any combination of species that can realize this value of m_{np} would be said to be 'damaged'. The mass inflow into plants can also be depressed below this reference value by simply eliminating vegetation, as is commonly done in 'developed' countries by concreting over the land surface. Massive withdrawal of circulating material, by soil erosion or cropping, can have a similar effect, and in this case the ecosystem can fairly be said to be 'damaged'. It seems unlikely that wild populations of herbivores, feeding on plants which have evolved defences against their depredations in non-productive periods, could ever produce 'damage' to an ecosystem in this sense. Probably, the ecosystems that supported the dinosaurs collapsed for other reasons. Human activities, however, can and do disrupt ecosystems, mainly because they involve flows of energy which are not found in natural ecosystems (Chapter 7).

6.15 Biomass redistribution

Representing an ecosystem in the manner of Fig. 6.1 is only valid under the restrictive conditions noted in Section 6.3. The ecosystem must have a boundary, within which material circulates, the masses in the boxes must refer to a particular element, and it must be an element which does not circulate through the atmosphere. If those conditions are satisfied, as they probably are at least for some elements in some ecosystems, then it follows that a finite amount of material has to be distributed among the various boxes representing the ecosystem. If material accumulates in one box, as a result of the organisms in that box being more than usually successful at biomass accumulation, then the material in one or more of the other boxes must be correspondingly reduced. Another way of looking at the changes in species composition, discussed in Section 6.13, is that they represent *biomass redistribution*. In a mature forest, a large fraction of the circulating stock may be incorporated into plant bodies, which leads to a depleted nutrient pool, and permits only modest densities of animal biomass. Destruction of trees and herbivore pressure has the effect of opening up the forest, so releasing material which can find its way via decomposers and the nutrient pool into smaller plants, and thence into animal biomass. Savannahs like those of East Africa have lower densities of plant biomass than forest, but can support higher animal biomass densities. It sometimes happens that the density of a particular species builds up to plague proportions, which implies that material has to be withdrawn from other compartments of the ecosystem, to create the additional biomass. For example, a plague of locusts exhibits an increase in the biomass density of locusts, and of animals that eat locusts, at the expense of a decrease in the biomass density of plants, and of herbivores that compete with locusts. One animal species in particular (our own) has in recent years achieved a world-wide degree of success at biomass accumulation, which appears to be unprecedented in the history of the biosphere. This has been accompanied by a precipitous decrease in the biomass of plants and of other animals. The present extreme prevalence of human biomass, which is still increasing, depends on a new and insecure type of material flow, and is the subject of the next chapter.

7. Ecosystems modified by human activities

7.1 Subsistence agriculture and the flow of materials

Human subsistence hunter–gatherer communities are not readily distinguishable from other generalist consumer species, from an ecological point of view. Subsistence farming has more of an impact on the local ecosystem, but remains a part of it. The subsistence farmer takes measures that divert the flow of material away from a number of naturally occurring plant species ('weeds'), into a small number of selected species ('crops'), which he then uses as a food supply, either directly for the human population, or for domestic animals which are in turn used as food. He also takes steps to prevent other primary consumers ('pests') from eating his crops. These measures are a very effective means of maximizing the inflow into human biomass (considered as a box in the manner of Fig. 6.1), and minimizing the outflow. The mass accumulation principle (Section 6.5) applies to human populations in just the same way as to those of other organisms, except that they have means at their disposal for continuing to accumulate mass, in circumstances under which other species would not be able to do so. As human biomass density rises thanks to agriculture, mass is no longer able to accumulate in large plant species (Section 6.15). Woodlands and forests open up, to be replaced by croplands and grasslands, dominated by smaller species of plants.

7.2 The change to market agriculture

Fundamental modifications to ecosystems follow from the change from subsistence to market agriculture, whereby agricultural products are removed from the area where they were grown, and consumed elsewhere. This is the step that makes most of the modern world function in a different manner from natural ecosystems. As long as goods were moved by human porters or animal transport, the rate at which material could be removed from an ecosystem remained modest enough that no drastic effects ensued. Major modifications of ecosystems became possible with the introduction of methods of transport powered by fossil fuels. Modern rail, road, and ship transport permits ecologically significant quantities of material to be moved very quickly over long distances. This depletes the circulating stock in the area where the crops were grown. If other forms of disturbance have the effect of accelerating soil erosion, dramatic 'dust-bowl' conditions may develop, as they did in parts of North America in the 1930s. Unlike the Tsavo dust-bowl, which was temporary, this type of disruption can exterminate most of the species in an ecosystem, and decimate the nutrient pool, so making recovery a matter of evolution rather than ecological adjustment. Formation of a new nutrient pool and a new set of species, to replace an ecosystem that has been wiped out, is a process requiring millennia rather than years or decades.

Modern market agriculture is based on balancing the flow of material in the manner of Fig. 7.1, which represents the flow of some element in a zone where crops are being grown and removed. The major difference from Fig. 6.1 is that only minor amounts of material now recirculate through the nutrient pool (thanks to the weeds and pests), while large amounts are withdrawn from the nutrient pool into plant or animal biomass, and then removed. In special circumstances, such a 'one-way' system could be maintained by an unusually large input from the mineral reserve

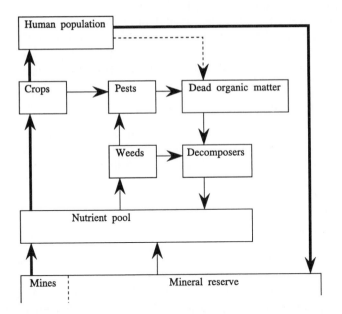

Fig 7.1 Flow diagram for an ecosystem dominated by market agriculture. The circulation of material that dominates Figure 6.1 has been largely replaced by a one-way flow from 'mines' (meaning fertilizer made from mineral sources), through crops and people, and out to the mineral reserve, without recirculation. Maintaining this unidirectional flow requires a large amount of energy, currently derived mainly from burning fossil fuel. Recirculation of human waste and dead bodies (dotted arrow) is probably small compared with the one-way flow (thick arrows) in most modern market agricultural systems. Natural circulation of materials is maintained only by such 'weeds' and 'pests' as escape destruction.

(m_{mn}). For instance, it is conceivable that a flood-plain such as that of the lower Nile, before the construction of the Aswan High Dam, could have received enough mineral input from flood-borne silt to sustain the annual removal of marketable quantities of plant material. Normally, however, the material removed by harvesting has to be replaced by mining minerals at some remote location. This requires the consumption of fossil fuel for transporting the material, and often also for converting it into a suitable chemical form. For example nitrogen has to be 'mined' from the atmosphere at a considerable cost in energy, before it can be transported as nitrogen fertilizer and used to replace the nitrogen removed in crops from agricultural land. Phosphorus in the form of guano used to be obtained by diverting part of the nutrient pool of a marine ecosystem off Peru to terrestrial agricultural ecosystems, but nowadays phosphorus fertilizer is also usually made from mineral sources. In the type of system shown in Fig. 7.1, the natural, local circulation of material has been replaced by a one-way flow out of the mineral reserve ('mines'), through crops and the human population. Although the material is ultimately returned to the mineral reserve, it does not recirculate. The mining of fertilizer is a one-time transfer of material. In some instances a modest part of the final waste is reused as fertilizer, but more often it is dumped via a sewage system into the ocean, where it is lost to terrestrial ecosystems.

7.3 Human biomass accumulation

As noted in Section 6.5, the mass accumulation principle is really the same as natural selection, so it is no surprise that people strive towards the same ends as all other organisms, past or present. Human populations strive to concentrate *all* of the circulating stock into their own biomass. Farmers everywhere feel deeply wronged if some weed or pest diverts even a minute proportion of the available flow of material and energy into a channel that does not end up in profitable human consumption. The new feature introduced by modern technology is that it is now actually possible to monopolize the flow of energy and material, and eliminate competing species, over vast areas of land.

As a yardstick to put human biomass density in perspective, we need an example of a dense population of large herbivores, in a productive natural habitat, such as the Serengeti wildebeest. According to Schaller (1972) the mass of an average migratory wildebeest is 108 kg, and the area over which the population roams is about 30 000 km², that is 3×10^{10} m². In the late 1980s the population number probably reached at least 1.4 million animals. On this conservative estimate, the biomass density, averaged over the whole range, works out at about $5 \cdot 0$ g m⁻². In the heyday of the 'elephant problem' in Tsavo National Park, Kenya, the biomass density of elephants there was similar. The habitat is less productive in Tsavo than in the Serengeti, but elephants are bigger than wildebeest, and so can maintain a similar biomass density on a smaller food supply (Section 6.9). Brown and Flavin (1988) estimate the world's human population in 1985 as about $4 \cdot 84 \times 10^9$ people, increasing by 1.7 per cent per year. To turn this into biomass, I shall use 50 kg as a guess at the average human mass. This might seem a low figure, but one has to remember that much of the world's population is undernourished, or young, or both. According to this estimate, the world's human biomass in 1985 was $2 \cdot 4 \times 10^{14}$ kg, increasing at $4 \cdot 1 \times 10^{12}$ kg per year. Stacey (1977) gives the total land surface of the world as $1 \cdot 48 \times 10^{14}$ m^2. On that basis, the average human biomass density, world-wide, is $1 \cdot 6$ g m^{-2}. Human biomass density, *world-wide*, comes within a factor of three of the Serengeti wildebeest.

The enormity of this result is difficult to comprehend. The figure of $1 \cdot 6$ g m^{-2} was obtained by dividing the estimated biomass of *one* species by the entire land surface of the earth, including Antarctica, Greenland, and the Sahara and Gobi deserts. The result is one-third of the biomass density of the Serengeti wildebeest, averaged over their miniscule 30 000 km^2 range. There are wildebeest outside the Serengeti, but not very many, and none outside Africa. Whether viewed in terms of population number ($1 \cdot 4$ million), or biomass ($1 \cdot 5 \times 10^8$ kg), this wildebeest population is comparable with the human population of a medium-sized city. Many countries, even small ones, have several cities that contain more human biomass than that. It is inconceivable that any one species in the past could ever have achieved a biomass density within orders of magnitude of 1 g m^{-2}, when averaged over the entire land surface of the planet. This is a situation entirely without precedent. The rate of increase of human biomass is apparently itself still increasing, which has to imply a corresponding rate of decrease in the amount of material available for all other forms of life (Section 6.15).

7.4 Local human biomass densities

Many an English family starts the day with American cornflakes, and finishes with New Zealand lamb, which makes it difficult to assign human consumers to an 'ecosystem' in the sense used in Chapter 6, having an identifiable boundary limiting the circulation of materials. The human biomass densities listed in Table 7.1 mostly do not represent recognizable ecosystems like the Serengeti. Biomass densities over 200 g m^{-2} are found in relatively small places like Hong Kong and Singapore, where the people make their living by commerce. Such figures have no relation to plant production, because virtually all of the food is imported. Countries like Britain and the Netherlands, with biomass densities over 10 g m^{-2} have to import a large percentage of their food (50 per cent or more) to keep their populations adequately fed. Bangladesh, which is situated on the Ganges

Table 7.1 Human biomass densities in different countries, assuming an average human mass of 50 kg; based on censuses or estimates between 1981 and 1984, listed in *The Times Atlas of the World* (7th Comprehensive Edition: 1985).

Country	Biomass density (g m^{-2})
Hong Kong	250
Singapore	200
Bangladesh	33
The Netherlands	18
Japan	16
United Kingdom	12
India	11
Haiti	9.1
Pakistan	5.5
China	5.3
Egypt	2.3
Kenya	1.6
USA	1.3
Sweden	0.93
Sudan	0.41
Iceland	0.12
Australia	0.10
Greenland	0.0012

Delta, and receives a huge annual input of nutrients eroded from the Himalayas, apparently does grow most of its own food, and manages to maintain a biomass density over 30 g m^{-2}. However, the population is said to be less well fed than the British or the Dutch. The Japanese maintain 16 g m^{-2} partly by importing food, and partly by harvesting (or mining) the world's oceans, in addition to their own land surface. Countries with benign climates, a high percentage of cultivable land surface, and modern agricultural systems seem (from Table 7.1) to be able to maintain human biomass densities up to 5 g m^{-2}, at least temporarily, without relying heavily on imported food, before signs of strain become apparent.

7.5 Biomass redistribution

The response of natural ecosystems to the continued rapid growth of human biomass can be readily understood in terms of the 'biomass redistribution' principle introduced in Chapter 6. When material is in short supply, but not to the extent that plant production is impaired, the balance of species composition automatically changes in favour of smaller species of both plants and animals. Nowadays, the materials of which new human bodies are made come partly from mined mineral fertilizer, but probably plant bodies are still the most important source. If that is the case, one would expect the increase in human biomass to be roughly matched by a corresponding decrease in plant biomass. Brown and Flavin's (1988) figures, noted above, suggest that the world's human biomass was increasing at about $4 \cdot 1 \times 10^{12}$ kg in 1985. As carbon makes up about 20 per cent of the elemental composition of human bodies, this would imply that carbon was being incorporated into new human biomass at a net rate of about $8 \cdot 2 \times 10^{11}$ kg per year. Crutzen and Andreae (1990) give a slightly later estimate of the world-wide rate at which carbon was being released by burning vegetation, as between 2×10^{12} and 5×10^{12} kg annually. Both of these estimates are quite rough, but their agreement within an order of magnitude is probably not a coincidence. Carbon is not in short supply, but other elements, needed to create new human biomass, are. The materials, out of which all those new human bodies are made, have to come from somewhere. The ash from burned plant bodies is a likely source for the major part of it.

As human populations increase, the forests disappear, to be replaced by open habitat with scattered cultivation. The plant biomass density is reduced further as remaining woody and herbaceous vegetation is eliminated and replaced by crops. In recent decades, the coming of scientific agriculture has made it possible to take the process a step further. By artificially breeding smaller and faster-growing crop plants, the intake rate of crop plants (R_p, in the terminology of Chapter 6), can be further increased. The 'green revolution' in some Asian countries was achieved partly by bringing even more land under cultivation − that is by diverting the flow of materials from the few remaining natural populations into human biomass − and partly by breeding smaller strains of rice and other crop plants. These 'improved' strains can produce the same amount of seed on a smaller plant body than their predecessors, so permitting a fixed rate of material flow through a smaller amount of plant biomass, that is, raising R_p. Of course, neither of these processes can be continued indefinitely. According to Brown (1990), the effect was most pronounced in the years 1950–1984, during which the world's grain output increased by the impressive factor of $2 \cdot 6$, but after 1984, there was no further increase through the rest of the 1980s.

7.6 Famine synchronization

When a population of large mammal herbivores in a productive habitat reaches a biomass density of the order of 10 g m^{-2}, the appearance of the vegetation becomes visibly threadbare. The depredations of smaller herbivores like mice or locusts become noticeable at lower biomass densities, because of the scaling relationship of Section 6.9. The maximum, sustainable herbivore biomass is determined by Equation 6.2, and when this limit is reached and passed, some mechanism has to come into operation, which decreases the rate of births, or increases the rate of deaths, or both. The adjustment often takes the form of a catastrophe, triggered by some fluctuation in the environment, like the drought that precipitated the Tsavo elephant deaths of 1959–60. Such episodes of mass mortality are known as population 'crashes'. Before the advent of market agriculture, human populations responded to climatic fluctuations in much the same way as other mammals. Famines were triggered by years of poor harvest at irregular but frequent intervals throughout the Middle Ages in Western Europe. The resulting human mortality from starvation is recorded in church registers and other documents.

Apart from these fluctuations, and occasional more widespread crashes such as that caused by the Black Death in 1347–50, the rate of increase of the human population was very gradual until the nineteenth century. With the coming of the railways, it became possible to move large quantities of food quickly over distances of a few tens or hundreds of kilometres, and thus relieve local famines. Most crop failures are comparatively local, seldom affecting the whole of even a small country like England. By importing food from another part of the country, where conditions were less severe, a local human population could survive one or more harvest failures without crashing. The elimination of local famines allowed the rate of increase of human biomass (and biomass density) to accelerate sharply in industrial countries, and this increase was accompanied by a progressive increase in the distance over which food could be economically transported, the amounts that could be moved, and the speed with which it could be done. In recent years there would have been widespread famine in the Soviet Union, if it had not been possible to import vast quantities of surplus grain cheaply from the USA. As a result of these measures, generations of westerners have lived out their lives without ever being confronted by the horror of famine, unlike our ancestors who experienced it every few years. There is a widespread belief that famine is a historical phenomenon only, but this is no more than a comfortable illusion. *Local* famine has been eliminated, but famine itself has not.

What has happened is that the build-up of human population density is now synchronized over large areas. The world of the Middle Ages was divided into many separate and independent ecosystems, whose human populations responded separately to the vagaries of their own local food supply. Nowadays, the entire North Temperate zone is one ecosystem as far as the human population is concerned, because surplus food can be transferred easily and cheaply from one part of it to another, to rectify local deficiencies. Serious famine is postponed until the *average* human biomass density over the whole vast area reaches a level which can no longer be sustained by crop production. This has permitted a long respite, since the days when local famines occurred two or three times per decade on average. Small famines occurring at frequent intervals have been exchanged for the potential for a very large famine, synchronized by the transport of food over the whole North Temperate zone, and perhaps beyond. This can only be averted if the biomass in the human box of Fig. 7.1 can somehow be restrained from building up beyond the point at which the available food supply is exceeded.

7.7 Controlling human biomass density

For the first time in the history of life on earth, it is now possible for one species (ours) to do what all organisms have been trying to do since life began. We can maintain the rate of flow of material into our own box, and restrict the flow out of it, until so little material is left to support other forms of life that the world ecosystem, or what remains of it, ceases to function. The world's animal biomass is already dominated by human biomass to an unprecedented degree, as noted in Section 7.3. As the imbalance continues to increase, so does the potential for a really spectacular crash, synchronized across continents and around the world. Human biomass has already built up far beyond the point at which it could be maintained by local circulation of material, as in natural ecosystems (Fig. 6.1). The one-way flow of Fig. 7.1 permits a higher level of biomass to be maintained, but only by using external sources of energy. These sources of energy might cease to be available for some reason, perhaps because supplies of fossil fuels are exhausted, or because the use of fossil or nuclear fuels produces side-effects in the form of pollution, or disturbances of the earth's heat balance. If the energy supply were cut off, the total animal biomass (dominated by human biomass) would have to revert to a much lower level, sustainable by local material circulation. The 'correction' would be more severe than a reversion to the previous status quo, because intensive agriculture has caused massive loss of material from terrestrial nutrient pools by soil erosion. Therefore the crash, when it comes, will be to a lower level of animal biomass density than was sustainable before the rise of market agriculture.

Individual people want to survive, and want their children to survive, as natural selection dictates. To a lumber-mill community, their own current income is more important than the survival of the forest, even if their own activities are obviously going to eliminate the source of their income in the near future. One should not dismiss this as stupidity. It is the result of thousands of millions of years of natural selection. *Every* organism has always done whatever has the short-term effect of increasing its own biomass. It is easy to point out that action is needed to reduce human biomass, but any such action would have to implemented by politicians, and they can only do what their people want them to do. Reducing human biomass means discouraging births, and encouraging deaths, which is contrary to the mass accumulation principle. Political, religious, or ethical systems cannot oppose the mass accumulation principle, because it is essentially the same as natural selection, and as old as life itself.

In much of Africa and Asia the human population has been increasing exponentially for some years. Although the dire consequences of this are well recognized, the World Bank, the United Nations, and the religious organizations have no choice but to do everything that they can to accelerate the process. Aid is readily available to put more land under cultivation, and to increase crop yields, so as to maximize the rate of mass flow (m_i) into the human population. Every effort is made to defeat disease, and to transport food to areas affected by local famines, so minimizing the rate of mass flow out of human biomass. To do otherwise would be impossible, because it would go against the mass accumulation principle, and thus against natural selection itself. The only foreseeable scenario for countries with this syndrome is a continuing exponential increase of human biomass, terminated in the not-too-distant future by a gigantic, and widely synchronized population crash.

7.8 Paying taxes direct to the ecosystem

There is a glimmer of hope in the 'developed' countries of Europe and North America, however. In several of them, human population growth has ceased, or nearly so. Although the reasons do not appear to have much to do with concerns felt by individual citizens about the future of the world ecosystem, it may nevertheless be possible to manipulate the economic system in a way that might stabilize the effect. In economically advanced countries, people think in terms of accumulating money rather than biomass, and this makes it possible to induce some very unnatural behaviour. For example, the American government has taken the view in recent years that production of food by farmers was too high in some areas (for economic, not ecological reasons), and has shown that it is possible to induce farmers *not* to cultivate part of their land. This is done by manipulating the tax system in such a way that the farmer makes more money by not cultivating the land than by cultivating it. If this principle were applied on a scientific, rather than an economic basis, it might be possible to restrain, or at least moderate, biomass accumulation. The method would be to survey each farm, and determine what proportion of the primary plant production is going into saleable crops, and what proportion into 'weeds and pests', that is, natural plant and animal populations. The farmer would only be taxed (in money) if the proportion going into crops exceeded some specified percentage. If, for the sake of argument, this percentage were fixed at 50 per cent, then the farmer would have to show that at least 50 per cent of the primary production was going into populations other than crops, in order to be excused paying tax on his profits. If too high a percentage was found to be going into crops, he would be taxed on his profits at a high enough rate to persuade him to conform. In effect, the farmer would pay his taxes in energy, direct to the ecosystem, instead of paying in money to the government. A system like this would make farmers into the most enthusiastic conservationists, as they would profit directly by preserving healthy natural populations of plants and animals on their land. The difficulty is that the main short-term losers would be politicians and bureaucrats, who unfortunately are the very people who would have to draft and pass the law.

7.9 Scientific priorities

A good many of the world's current problems (perhaps all) can be traced to the fact that the world's animal biomass is dominated by one species (us), to an excessive degree, and the imbalance is daily becoming more extreme. In view of that, one should consider the merits of different types of applied research in the light of its probable effect on the future course of human biomass. The largest effort and expenditure in the area of 'biomedical' research is directed towards artificially prolonging human lives, which has been a major factor in bringing the present biomass imbalance into existence. The next largest component is agricultural research, which is directed towards maximizing the mass inflow to the human population, at the expense of other forms of life. Proponents of these areas of research get their funds from politicians by claiming that their purpose is to 'save lives' or 'alleviate suffering'. Because of the close connection between biomass accumulation and natural selection, noted above, these phrases are understood by politicians and the public alike to mean any activity that promotes increased human biomass accumulation.

It is time to begin asking whether 'saving lives' is still a defensible priority in a world already overrun by people. The vast research effort currently devoted to medical science is out of proportion to any benefits that may be expected to follow from it. It is more urgent to find out whether there are any options left, which offer a chance of stabilizing human biomass in relation to other forms of life. Politicians and economists will continue to generate solutions to short-term problems, but we lack a sufficient theoretical basis to determine whether these schemes are really as beneficial as their proponents invariably claim, or whether they will cause worse problems than they solve. A better understanding is needed of the way in which both natural and artificial ecosystems work. These are newtonian systems, and I suggest that a newtonian approach is the best way to formulate the large-scale laws that govern them, and in particular the rules that determine how material is distributed. The emphasis in biological research funding needs to be shifted away from medicine and agriculture, and towards fundamental studies of ecosystem functioning.

References

Alexander, R. McN. (1980). Optimum walking techniques for quadrupeds and bipeds. *Journal of Zoology (London)*, **192**, 97–117.

Alexander, R. McN. and Jayes, A. S. (1983). A dynamic similarity hypothesis for the gaits of quadrupedal mammals. *Journal of Zoology (London)*, **201**, 135–52.

Alexander, R. McN., Langman, V. A., and Jayes, A. S. (1977). Fast locomotion of some African ungulates. *Journal of Zoology (London)*, **183**, 291–300.

Anderson, J. D. (1984). *Fundamentals of aerodynamics*. McGraw Hill, New York.

Bennet-Clark, H. C. and Lucey, E. C. A. (1967). The jump of the flea: a study of the energetics and a model of the mechanism. *Journal of Experimental Biology*, **47**, 59–76.

Boettiger, E. G. (1957). Triggering of the contractile process in insect fibrillar muscle. In *Physiological triggers* (ed. T. H. Bullock), pp. 103–16. American Physiological Society.

Brown, L. R. (1990). The illusion of progress. In *State of the World 1990* (ed. L. Starke), pp. 3–16. Worldwatch Institute (Norton), New York.

Brown, L. R. and Flavin, C. (1988). The Earth's vital signs. In *State of the World 1988* (ed. L. Starke), pp. 3–21. Worldwatch Institute (Norton), New York.

Campbell, K. E. and Tonni, E. P. (1983). Size and locomotion in teratorns (Aves: Teratornithidae). *Auk*, **100**, 390–403.

Carey, F. G. (1973). Fishes with warm bodies. *Scientific American*, **228**, 36–44.

Carey, S. W. (1976). *The Expanding Earth*. Elsevier, Amsterdam.

Crutzen, P. J. and Andreae, M. O. (1990). Biomass burning in the tropics: impact on atmospheric chemistry and biogeochemical cycles. *Science*, **250**, 1669–78.

Ellington, C. P. (1985). Power and efficiency of insect flight muscle. *Journal of Experimental Biology*, **115**, 293–304.

Fung, Y. C. (1981). *Biomechanics*. Springer, New York.

Gnaiger, E. (1989). Physiological calorimetry: heat flux, metabolic flux, entropy and power. *Thermochimica Acta*, **151**, 23–34.

Heglund, N. C. and Taylor, C. R. (1988). Speed, stride frequency and energy cost per stride: how do they change with body size and gait? *Journal of Experimental Biology*, **138**, 301–18.

Hill, A. V. (1938). The heat of shortening and the dynamic constants of muscle. *Proceedings of the Royal Society, Series B*, **126**, 136–95.

Hill, A. V. (1950). The dimensions of animals and their muscular dynamics. *Science Progress*, **38**, 209–30.

Hofman, M. A. (1988). Allometric scaling in palaeontology: a critical survey. *Human Evolution*, **3**, 177–88.

Huxley, A. F. (1957). Muscle structure and theories of contraction. *Progress in Biophysics and Biophysical Chemistry*, **7**, 255–318.

Huxley, H. E. (1985). The crossbridge mechanism of muscular contraction and its implications. *Journal of Experimental Biology*, **115**, 17–30.

Johnston, I. A. (1985). Sustained force development: specializations and variation among the vertebrates. *Journal of Experimental Biology*, **115**, 239–51.

Kirkpatrick, S. J. (1990). The moment of inertia of bird wings. *Journal of Experimental Biology*, **151**, 489–94.

Klein, H. A. (1974). *The science of measurement. A historical survey.* Simon and Schuster, New York (Dover edition, 1988).

Mandelbrot, B. B. (1967). How long is the coast of Britain? Statistical self-similarity and fractional dimension. *Science*, **155**, 636–8.

Mandelbrot, B. B. (1983). *The fractal geometry of nature.* Freeman, New York.

McMahon, T. A. (1984). *Muscles, reflexes and locomotion.* Princeton University Press, Princeton, New Jersey.

McMahon, T. A. and Bonner, J. T. (1983). *On size and life.* Scientific American Books, New York.

McNab, B. K. (1983). Energetics, body size, and the limits to endothermy. *Journal of Zoology (London)*, **199**, 1–29.

Moreau, R. E. (1966). *The bird faunas of Africa and its islands.* Academic Press, New York.

Pennycuick, C. J. (1975). On the running of the gnu (*Connochaetes taurinus*) and other animals. *Journal of Experimental Biology*, **63**, 775–99.

Pennycuick, C. J. (1982). The flight of petrels and albatrosses (Procellariiformes), observed in South Georgia and its vicinity. *Philosophical Transactions of the Royal Society, Series B*, **300**, 75–106.

Pennycuick, C. J. (1987). Cost of transport and performance number, on earth and other planets. In *Comparative physiology: life in water and on land* (ed. P. Dejours, L. Bolis, C. R. Taylor, and E. R. Weibel), pp. 371–86. Liviana Press, Padova, Italy.

Pennycuick, C. J. (1988). *Conversion factors: SI units and many others.* University of Chicago Press.

Pennycuick, C. J. (1989). *Bird flight performance: a practical calculation manual.* Oxford University Press.

Pennycuick, C. J. (1990). Predicting wingbeat frequency and wavelength of birds. *Journal of Experimental Biology*, **150**, 171–85.

Pennycuick, C. J. (1992). Adapting skeletal muscles to be efficient. In *Efficiency, economy and related concepts in comparative animal physiology* (ed. R. W. Blake). Cambridge University Press.

Pennycuick, C. J. and Kline, N. C. (1986). Units of measurement for fractal extent, applied to the coastal distribution of bald eagle nests in the Aleutian Islands, Alaska. *Oecologia (Berlin)*, **68**, 254–8.

Pennycuick, C. J. and Rezende M. A. (1984). The specific power output of aerobic muscle, related to the power density of mitochondria. *Journal of Experimental Biology*, **108**, 377–92.

Rayner, J. M. V. (1985). Linear relations in biomechanics: the statistics of scaling. *Journal of Zoology (London)*, **206**, 415–39.

Richardson, L. F. (1961). The problem of contiguity: an appendix of statistics of deadly quarrels. *General Systems Yearbook*, **6**, 139–87.

Schaller, G. B. (1972). *The Serengeti Lion*. University of Chicago Press.

Schmidt-Nielsen, K. (1972). Locomotion: energy cost of flying, swimming and running. *Science*, **177**, 222–8.

Sinclair, A. R. E. and Norton-Griffiths, M. (1979). *Serengeti: dynamics of an ecosystem*. University of Chicago Press.

Stacey, F. D. (1977). *Physics of the earth*, (2nd edn). Wiley, New York.

Stevenson, R. D. and Josephson, R. K. (1990). Effects of operating frequency and temperature on mechanical power output from moth flight muscle. *Journal of Experimental Biology*, **149**, 61–78.

Western, D. (1980). Linking the ecology of past and present mammal communities. In *Fossils in the making* (ed. A. K. Behrensmeyer and A. P. Hill), pp. 41–54. University of Chicago Press.

White, D. C. S. and Thorson, J. (1975). *The kinetics of muscle contraction*. Pergamon, Oxford.

Index